Advanced Manufacturing and Processing Technology

Manufacturing Design and Technology Series

Series Editor:
J. Paulo Davim

This series will publish high-quality references and advanced textbooks in the broad area of manufacturing design and technology, with a special focus on sustainability in manufacturing. Books in the series should find a balance between academic research and industrial application. This series targets academics and practicing engineers working on topics in materials science, mechanical engineering, industrial engineering, systems engineering, and environmental engineering as related to manufacturing systems, as well as professions in manufacturing design.

Drills
Science and Technology of Advanced Operations
Viktor P. Astakhov

Technological Challenges and Management
Matching Human and Business Needs
Edited by Carolina Machado and J. Paulo Davim

Advanced Machining Processes
Innovative Modeling Techniques
Edited by Angelos P. Markopoulos and J. Paulo Davim

Management and Technological Challenges in the Digital Age
Edited by Pedro Novo Melo and Carolina Machado

Machining of Light Alloys
Aluminum, Titanium, and Magnesium
Edited by Diego Carou and J. Paulo Davim

Additive Manufacturing
Applications and Innovations
Edited by Rupinder Singh and J. Paulo Davim

For more information about this series, please visit: https://www.routledge.com/Manufacturing-Design-and-Technology/book-series/CRCMANDESTEC

Advanced Manufacturing and Processing Technology

Edited by
Chander Prakash, Sunpreet Singh, and
J. Paulo Davim

CRC Press is an imprint of the
Taylor & Francis Group, an **informa** business

First edition published 2021
by CRC Press
6000 Broken Sound Parkway NW, Suite 300, Boca Raton, FL 33487-2742

and by CRC Press
2 Park Square, Milton Park, Abingdon, Oxon, OX14 4RN

© 2021 Taylor & Francis Group, LLC

CRC Press is an imprint of Taylor & Francis Group, LLC

Reasonable efforts have been made to publish reliable data and information, but the author and publisher cannot assume responsibility for the validity of all materials or the consequences of their use. The authors and publishers have attempted to trace the copyright holders of all material reproduced in this publication and apologize to copyright holders if permission to publish in this form has not been obtained. If any copyright material has not been acknowledged please write and let us know so we may rectify in any future reprint.

Except as permitted under U.S. Copyright Law, no part of this book may be reprinted, reproduced, transmitted, or utilized in any form by any electronic, mechanical, or other means, now known or hereafter invented, including photocopying, microfilming, and recording, or in any information storage or retrieval system, without written permission from the publishers.

For permission to photocopy or use material electronically from this work, access www.copyright. com or contact the Copyright Clearance Center, Inc. (CCC), 222 Rosewood Drive, Danvers, MA 01923, 978-750-8400. For works that are not available on CCC please contact mpkbookspermissions@ tandf.co.uk

Trademark notice: Product or corporate names may be trademarks or registered trademarks, and are used only for identification and explanation without intent to infringe.

Library of Congress Cataloging-in-Publication Data
Names: Prakash, Chander, editor. | Singh, Sunpreet, editor. |
Davim, J. Paulo, editor.
Title: Advanced manufacturing and processing technology / edited by Chander
Prakash, Sunpreet Singh, and J. Paulo Davim.
Description: First edition. | Boca Raton, FL : CRC Press, [2021] |
Series: Manufacturing design and technology | Includes bibliographical
references and index.
Identifiers: LCCN 2020018666 (print) | LCCN 2020018667 (ebook) |
ISBN 9780367275129 (hardback) | ISBN 9780429298042 (ebook)
Subjects: LCSH: Manufacturing processes. | Materials science.
Classification: LCC TS183 .A37 2021 (print) | LCC TS183 (ebook) |
DDC 670—dc23
LC record available at https://lccn.loc.gov/2020018666
LC ebook record available at https://lccn.loc.gov/2020018667

ISBN: 978-0-367-27512-9 (hbk)
ISBN: 978-0-429-29804-2 (ebk)

Typeset in Times
by codeMantra

Contents

Preface vii

Editors ix

Contributors xi

Chapter 1   A Critical Review on the Machining of Engineering Materials
by Die-Sinking EDM 1

*P. Sreeraj, S. Thirumalai Kumaran, M. Uthayakumar,
and S. Suresh Kumar*

Chapter 2   Optimization of Machining Parameters of High-Speed Toolpath
to Achieve Minimum Cycle Time for Ti-6Al-4V 23

*Ganesh Kakandikar, Rahul Dhage, Omkar Kulkarni, and
Vilas Nandedkar*

Chapter 3   A Review of Machinability Aspects of Difficult-to-Cut
Materials Using Microtexture Patterns 45

Rahul Sharma, Swastik Pradhan, and Ravi Nathuram Bathe

Chapter 4   Micromachining 67

Venkatasreenivasula Reddy Perla and Rathanraj K.J.

Chapter 5   A Review Study on Miniaturization—A Boon or Curse 111

Ankit Sharma, Vivek Jain, Dheeraj Gupta, and Atul Babbar

Chapter 6   A Comprehensive Review on Similar and Dissimilar Metal
Joints by Friction Welding 133

D. Saravanakumar, M. Uthayakumar, and S. Thirumalai Kumaran

Chapter 7   3D Bioprinting in Pharmaceuticals, Medicine, and Tissue
Engineering Applications 147

*Atul Babbar, Vivek Jain, Dheeraj Gupta, Chander Prakash,
Sunpreet Singh, and Ankit Sharma*

v

vi Contents

Chapter 8 Investigating on the Lapping and Polishing Process of Cylindrical Rollers 163

Duc-Nam Nguyen, Ngoc Le Chau, and Thanh-Phong Dao

Chapter 9 NiTi Thin-Film Shape Memory Alloys and Their Industrial Application 185

Ajit Behera, Patitapabana Parida, and Aditya Kumar

Chapter 10 Carbon Fibers: Surface Modification Strategies and Biomedical Applications... 207

Suneev Anil Bansal, Javad Karimi, Amrinder Pal Singh, and Suresh Kumar

Index.. 225

Preface

The goal of this book is to provide technical insights on recent innovations in the different classes of advanced manufacturing and processing technologies available in modern manufacturing sectors. We have detailed and collected information on working principle, process mechanism, salient features, and unique applications of various advanced manufacturing techniques. This book aims to present trends and recent research breakthroughs in the field of manufacturing processes, optimization and process planning, metal forming, integrated manufacturing systems, nonconventional machining, additive manufacturing, robotics, process control, measurement, and quality control.

We cover important research papers and review articles illustrating the latest developments in advanced manufacturing and processing technologies. We also cover contributions dealing with novel materials and advanced manufacturing technologies. The original research work on the manufacturing of different types of ferrous and nonferrous alloys, composite materials, and polymers, their mechanical, tribological, and metallurgical performances, surface engineering, thermodynamics and kinetics, mathematical modeling, and other allied domains of manufacturing have been duly discussed. The content of the book bridges the gap between modern engineering practices and engineering innovations and, therefore, provides an outstanding forum for research efforts covering applications-based research topics relevant to manufacturing processes.

Editors

Chander Prakash is Associate Professor at the School of Mechanical Engineering, Lovely Professional University, Jalandhar, India. He received a Ph.D. in Mechanical Engineering from Panjab University, Chandigarh, India. His areas of research is biomaterials, rapid prototyping and 3D printing, advanced manufacturing, modeling, simulation, and optimization. He has more than 11 years of teaching experience and 6 years of research experience. He has contributed extensively to titanium- and magnesium-based implant literature with publications appearing in *Surface and Coating Technology*, *Materials and Manufacturing Processes*, *Journal of Materials Engineering and Performance*, *Journal of Mechanical Science and Technology*, *Nanoscience and Nanotechnology Letters*, and *Proceedings of the Institution of Mechanical Engineers*, *Part B: Journal of Engineering Manufacture*. He has authored 150 research papers and 30 book chapters. He is also an editor of 15 Books: He is also a guest editor of two journals: Special Issue on "Metrology in Materials and Advanced Manufacturing," *Measurement and Control* (SCI indexed) and Special Issue on "Nano-Composites and Smart Materials: Design, Processing, Manufacturing and Applications" of *Advanced Composites Letters*.

Sunpreet Singh is researcher in NUS Nanoscience & Nanotechnology Initiative (NUSNNI). He received a Ph.D. in Mechanical Engineering from Guru Nanak Dev Engineering College, Ludhiana, India. His area of research is additive manufacturing and application of 3D printing for development of new biomaterials for clinical applications. He has contributed extensively to the subject of additive manufacturing with publications appearing in *Journal of Manufacturing Processes*, *Composite Part: B*, *Rapid Prototyping Journal*, *Journal of Mechanical Science and Technology*, *Measurement*, *International Journal of Advance Manufacturing Technology*, and *Journal of Cleaner Production*. He has authored 10 book chapters and monographs. He is working in joint collaboration with Prof. Seeram Ramakrishna, NUS Nanoscience & Nanotechnology Initiative and Prof. Rupinder Singh, Manufacturing Research Lab, GNDEC, Ludhiana. He is also an editor of three books: *Current Trends in Bio-manufacturing*, Springer Series in Advanced Manufacturing, Springer International Publishing AG, Gewerbestrasse 11, 6330 Cham, Switzerland., December 2018; *3D Printing in Biomedical Engineering*, Book series Materials Horizons: From Nature to Nanomaterials, Springer International Publishing AG, Gewerbestrasse 11, 6330 Cham, Switzerland, August 2019; and *Biomaterials in Orthopaedics and Bone Regeneration - Design and Synthesis*, Book series: Materials Horizons: From Nature to Nanomaterials, Springer International Publishing AG, Gewerbestrasse 11, 6330 Cham, Switzerland, March 2019. He is also Guest Editor of three journals: Special Issue on "Functional Materials and Advanced Manufacturing," Facta Universitatis, Series: Mechanical Engineering (Scopus Index), Materials Science Forum (Scopus Index), and Special Issue on "Metrology in Materials and Advanced Manufacturing," *Measurement and Control* (SCI indexed).

J. Paulo Davim received a Ph.D. in Mechanical Engineering in 1997, a M.Sc. degree in Mechanical Engineering (materials and manufacturing processes) in 1991, a Mechanical Engineering degree (five years) in 1986 from the University of Porto (FEUP), the Aggregate title (Full Habilitation) from the University of Coimbra in 2005, and a D.Sc. from London Metropolitan University in 2013. He is Senior Chartered Engineer by the Portuguese Institution of Engineers with an MBA and Specialist title in Engineering and Industrial Management. He is also Eur Ing by FEANI-Brussels and Fellow (FIET) by IET-London. Currently, he is a professor at the Department of Mechanical Engineering of the University of Aveiro, Portugal. He has more than 30 years of teaching and research experience in manufacturing, materials, and mechanical and industrial engineering, with special emphasis in machining and tribology. He has also interest in management, engineering education, and higher education sustainability. He has guided many postdoc, Ph.D. and master's students as well as coordinated and participated in several financed research projects. He has received several scientific awards. He has worked as evaluator of projects for the European Research Council (ERC) and other international research agencies as well as examiner of Ph.D. candidates for many universities in different countries. He is the editor-in-chief of several international journals, a guest editor of journals, books series, and a scientific advisor for many international journals and conferences. Presently, he is an editorial board member of 30 international journals and acts as reviewer for more than 100 prestigious Web of Science journals. He has also published as editor (and co-editor) of more than 100 books and as author (and co-author) of more than 10 books, 80 book chapters, and 400 articles in journals and conferences (more than 250 articles in journals indexed in Web of Science core collection/h-index 49+/7000+ citations, SCOPUS/h-index 56+/10000+ citations, Google Scholar/h-index 70+/16000+).

Contributors

Atul Babbar
Mechanical Engineering Department
Thapar Institute of Engineering and
  Technology
Patiala, India

Suneev Anil Bansal
Department of Mechanical Engineering
MAIT, Maharaja Agrasen University
Barotiwala, India

Ravi Nathuram Bathe
International Advanced Research
  Centre for Powder Metallurgy and
  New Materials (ARCI)
Hyderabad, India

Ajit Behera
Department of Metallurgical and
  Materials Engineering
NIT—Rourkela
Rourkela, India

Ngoc Le Chau
Faculty of Mechanical Engineering
Industrial University of Ho Chi
  Minh City
Ho Chi Minh City, Vietnam

Thanh-Phong Dao
Division of Computational
  Mechatronics
Institute for Computational Science
Ton Duc Thang University
Ho Chi Minh City, Vietnam
Faculty of Electrical & Electronics
  Engineering
Ton Duc Thang University
Ho Chi Minh City, Vietnam

Rahul Dhage
Unicorn CNC Technologies,
  Hadapsar
Pune, India

Dheeraj Gupta
Mechanical Engineering
  Department
Thapar Institute of Engineering and
  Technology
Patiala, India

Vivek Jain
Mechanical Engineering Department
Thapar University
Patiala, India

Ganesh Kakandikar
School of Mechanical Engineering
Dr. V. D. Karad MIT World Peace
  University
Pune, India

Javad Karimi
Department of Biology, Faculty of
  Science
Shiraz University
Shiraz, Iran

Omkar Kulkarni
School of Mechanical Engineering
Dr. V. D. Karad MIT World Peace
  University
Pune, India

Aditya Kumar
Department of Chemical Engineering
ISM—Dhanbad
Dhanbad, India

Suresh Kumar
Department of Applied Sciences
UIET, Panjab University
Chandigarh, India

Vilas Nandedkar
Production Engineering Department
SGGS Institute of Engineering and
    Technology
Nanded, India

Duc-Nam Nguyen
Faculty of Mechanical Engineering
Industrial University of Ho Chi
    Minh City
Ho Chi Minh City, Vietnam

Patitapabana Parida
Department of Metallurgical and
    Materials Engineering
NIT—Rourkela
Rourkela, India

Venkatasreenivasula Reddy Perla
Research Scholar

Swastik Pradhan
School of Mechanical Engineering
Lovely Professional University
Phagwara, India

Chander Prakash
Chitkara College of Applied
    Engineering
Chitkara University
Patiala, India

Rathanraj K.J.
Department of Industrial Engineering
    and Management
B.M.S. College of Engineering
Bengaluru, India

D. Saravanakumar
Kalasalingam Academy of Research
    and Education
Krishnankoil, India

Ankit Sharma
Chitkara College of Applied
    Engineering
Chitkara University
Punjab, India
School of Mechanical Engineering
Lovely Professional University
Phagwara, India

Rahul Sharma
School of Mechanical Engineering
Lovely Professional University
Phagwara, India

Amrinder Pal Singh
Department of Mechanical Engineering
UIET, Panjab University
Chandigarh, India

Sunpreet Singh
Chitkara College of Applied
    Engineering
Chitkara University
Patiala, India

Sreeraj P.
Kalasalingam University
Krishnankoil, India

Suresh Kumar S.
Kalasalingam University
Krishnankoil, India

S. Thirumalai Kumaran
Kalasalingam Academy of Research
    and Education
Krishnankoil, India

M. Uthayakumar
Kalasalingam Academy of Research
    and Education
Krishnankoil, India

1 A Critical Review on the Machining of Engineering Materials by Die-Sinking EDM

P. Sreeraj, S. Thirumalai Kumaran,
M. Uthayakumar, and S. Suresh Kumar
Kalasalingam Academy of Research and Education

CONTENTS

1.1 Introduction 1
1.2 Researches on Electrical Discharge Machining Process................................. 2
    1.2.1 Researches on Enhancement of Tool Wear 2
    1.2.2 Optimization of Process Parameters 3
    1.2.3 Selection of Electrode Material 4
    1.2.4 Multispark Erosion Studies... 6
    1.2.5 Selection of Optimized Pulse Duration 7
    1.2.6 Vibratory Tool and Workpiece 8
    1.2.7 Introducing Servo Control Mechanism 9
    1.2.8 Magnetic Field–Based Electrical Discharge Machining...................... 9
    1.2.9 Special Tools... 10
    1.2.10 CNC-Controlled Electrical Discharge Machining 11
    1.2.11 Selection of Dielectric Medium... 11
1.3 Application of Electrical Discharge Machining for Biomaterials................. 13
1.4 Conclusions... 13
References... 14

1.1 INTRODUCTION

In automotive and aerospace industries for producing complicated shapes of high hardness and toughness, electrical discharge machining (EDM) has been commonly used [1]. Conversion of electrical energy to thermal energy is taking place by means of discrete sparks through separating dielectric medium, which causes material removal. Plasma channel is generated between electrodes, and this high temperature around 20,000°C melts the electrodes, which cause the material removal. When the plasma channel breaks down, materials are flushed away by means of circulating die electric fluid, which actually causes material removal.

Nowadays, EDM is employed as a standard technique for machining of hardened new-generation alloys where traditional machining techniques have proved to be inefficient. Frequent changes of electrodes are required because of rapid tool wear. Machining accuracy of process is also influenced by tool wear, and hence, cost of replacement of the tool increases. In EDM, tool and workpiece have been considered as electrodes, and erosion occurs on both the electrodes. Electrode wear can be classified into four types such as volumetric, corner, end, and side wears [2]. Among them, corner wear has significant influence on affecting the accuracy of machining.

Minimization of electrode wear is one of the research areas to be concentrated in EDM. Considerable research has been done in this area for the minimization of tool wear rate (TWR). Likewise, improvement of material removal rate (MRR), that is, maximization of machining rate and improvement of surface quality (SQ), is also a major challenge for researchers. When an electrode is connected to negative polarity, MRR found to be influenced by recast layer creation. Regarding quality of each prepared surface EDM, it consists of three layers such as recast or white overlay, heat-affected zone, and original substrate material. For improving SQ in machining through research, these three layers are required. Quality of machined surface depends on white layer because it is the topmost layer exposed to atmosphere [3].

This chapter concentrates on standard techniques suggested and contributed by different authors for the reduction of tool erosion rate, maximization of material expulsion rate, and SQ. Furthermore, scope for future work has been proposed, and it may be helpful for those who conduct studies in the field of EDM.

1.2 RESEARCHES ON ELECTRICAL DISCHARGE MACHINING PROCESS

A complete analysis of research activities conducted in the past decades in die-sinking EDM process has been discussed in this section. There are quantitatively good advantages in EDM such as high precision production of mirror image. In spite of these advantages, EDM has many disadvantages such as formation of surface cracks, metallurgical changes in surface and subsurface regions, formation of recast layers, and formation of heat-affected zones. Researchers have devised many methods to overcome these deficiencies of EDM. The following sections describe the methodologies and strategies adopted in EDM process for enhancing the efficiency and quality of machining.

1.2.1 RESEARCHES ON ENHANCEMENT OF TOOL WEAR

Melting of electrodes, due to powerful dislodgement of superheated electrodes at the end of each pulse, causes material removal in EDM. In this development, both tool and workpiece undergo some surface changes because of this melting phenomenon. Many researchers have studied about workpiece modification, abatement of tool wear, and maximization of MRR. Equally important issue is reduction of tool wear. Soni et al. [4] have noticed that appreciable quantity of material migrates from workpiece to tool by forming a protective layer, which causes the reduction of tool wear. Marafona et al. [5] have proved that reduction of tool wear is achieved by a protective thin film of carbon formed on the surface of the tool. Researchers have also shown that this carbon film has considerable effect on reduction of tool wear but has no role on MRR. Marafona [6] has found that because of shift of carbon from circulating dielectric medium, a black surface on tool (W/Cu) is produced while machining the workpiece BS 4695 D2, and the presence of iron, chromium, vanadium, and molybdenum in black layer is also found. These elements present in tool surface form carbon equivalent, which is the main reason for the reduction of tool wear.

Mohri et al. [7,8] proved that in the beginning of machining, electrode wear is more at the edge than flat portion, and it increases in longitudinal direction. In the edge portion, proper deposit of carbon does not take place from die electric during sparking process, and it is the main reason for more wear in the edge portion. Researchers have reported that by providing proper protective layer on tool surface and by using dry medium, the tool wear can be minimized considerably. The ultimate aim of the new researchers is to obtain zero wear electrodes.

In order to achieve improvement in mechanical properties, a recast film has been on EDMed surface of AISI H13 tool steel. Amorim et al. [9] have conducted experiments with electrolyte copper as electrode and hydrocarbon oil as die electric with suspended molybdenum (Mo). The results have indicated that a coating has been formed on the surface and it has improved hardness and other mechanical properties.

1.2.2 Optimization of Process Parameters

Input parameters influence various aspects such as material depletion rate, surface irregularities, tool erosion rate, radial over cut, cutting width, cutting agility, and crater width. By controlling and optimizing major process parameters, the quality of machining can be improved [10]. EDM of Al–10% SiCp composite has been carried out by Vinothkumar et al. [11] for investigating and comparing influence of cryogenically cooled copper electrode and conventional electrode. For cryogenic cooling, liquid nitrogen (LN_2) is used. Grey relational analysis (GRA) has been used for optimizing the process framework, and result shows that cryogenic cooling gives increased efficiency. Majumder [12] has conducted experiments on AISI 316LN stainless steel for optimizing the process parameters in EDM process. Researchers have made fuzzy model to provide fitness function and finally used particle swarm optimization (PSO) for optimizing multiobjective function. Teimouri and Baseri [13] presented a novel model for removing the debris from tool–workpiece interface by using a rotary tool assisted with rotary magnetic field.

Shabgard et al. [14] investigated the influence of process parameters of AISI H13 tool steel work piece and AISI H13 steel tool. The research has shown that the controlling of processes parameters can considerably reduce TWR. The significance of operational parameters on machining performances of EDM has been studied by Huang et al. [15]. Central composite design has been used with different input process responses and two output parameters [16]. Analysis of variance has been utilized for investigating most important process measures in this process. In the aforementioned work, different flushing methods have been used for conducting experiments.

GRA combined with principal component analysis (PCA) has been applied by Jadish et al. [17] to evaluate and optimize the EDM performance characteristics. PCA-weighted components corresponding to the performance measures are first evaluated, and then by applying GRA, the process has been optimized. Most significant factor among the process parameters is determined by analysis of variance (ANOVA). Instead of conventional Response Surface Methodology (RSM) model, Xuan-Phuong Dang [18] has used Kriging model for creating nonlinear equations related to MRR and tool wear with standard input parameters in CNC EDM of P20 steel. Then, PSO has been used for optimizing this multiobjective optimization problem. It is also proved that this method is successful for other types of applications. A new approach of optimization known as Neuro-Grey method of optimization has been presented by Panda [19] and Tripathy and Tripathy [20] in EDM. In this process, commercial grade oil is used as dielectric, and SiC powder is used as an additive. This attempt has been proved to be very successful. Manivannan and Pradeep Kumar [21] have conducted drilling of microhole in AISI 304 stainless steel using hollow brass tube as electrode. For making cooling effect, liquid nitrogen is supplied to machining zone. Optimization of process has been carried out by TOSIS method.

Aliakbari and Baseri [22] have used rotary copper tool for conducting EDM experiments. They used conventional Taguchi method for optimizing the process parameters. Marafona and Jo [23] have attempted to create a response surface model for studying the impact of material hardness on MRR and surface roughness (SR). Ghamdi and Taylan [24] have shown that this method is advantageous and useful for modeling EDM. From the aforementioned research works, it is proved that by optimizing the process parameters, considerable improvement has been achieved during machining by die-sinking EDM. Newer mathematical models are yet to be developed for solving multiobjective problems in EDM.

1.2.3 Selection of Electrode Material

Appropriate selection of electrode is a herculean effort in EDM activity. Various measures such as MRR, tool erosion rate, and SR are to be considered while selecting the material for an electrode. Since the electric current is used as the cutting mechanism in EDM process, a tool with higher electrical conductivity or resistivity is normally preferred. Usually, the electrodes are classified into two, and they are metallic and graphite. Graphite has high melting point which converts directly from solid to gas at higher temperature. This makes graphite an ideal tool for machining. As EDM is a thermoelectric process, the tool should have sufficient melting point and low wear ratio. Though the mechanical properties do not affect the EDM because of no metal for tool contact, the hardness, grain size, and the mode of preparation of electrodes are very crucial. While selecting the electrode for machining, the TWR, MRR, and SQ to be obtained after machining of workpiece should be considered

In EDM process, higher conductivity and low resistivity are the general requirements of electrode. It is the tool electrode that transmits the electrical current to the workpiece separated by a circulating dielectric medium. Other important factors considered while selecting the electrodes are workpiece material, MRR, electrode resistance to wear, workpiece roughness, tool electrode material machinability, and cost of the electrode.

Different types of stainless steels with high corrosion resistances such as 316 L and 17-4 PH have been EDMed by Gopalakannan et al. [25] by selecting copper, copper–tungsten, and graphite electrodes. It was shown by researchers that copper electrode with high MRR achieved, but copper–tungsten electrode results indicate lower electrode wear, good surface finish, and better dimensional accuracy. Torres et al. [26] investigated machinabilty of INCONEL 600 by providing graphite electrodes in die-sinking EDM process. In this machining, dielectric fluid is the mineral oil because it is most widely used in EDM. It is evident from analysis that higher MRR is achieved when negative polarity is used, whereas lower electrode wear and good SR were obtained, when a positive polarity is used.

Experiments have been conducted for drilling of holes with conventional EDM and dry EDM in AISI D2 steel using copper electrode tool in EDM, and it has been proposed without central core by Pragadish and Kumar [27]. They have argued that these methods will cause effective removal of debris. While comparing the results, it is shown that good surface characteristics have been achieved in machined surface under dry EDM process when copper electrode tool is used. A special designed tool has been used for flow of dielectric oxygen.

Rotary tool has been considered in machining of high-chromium high-carbon die steel using argon gas–assisted EDM (AGAEDM) by Singh et al. [28]. In EDM process, workpiece along with tool experiences an exorbitant heating in the vicinity of plasma channel. In order to overcome this issue, an adequate attention should be given in the selection of tool material, electrode polarity, and process parameters. According to Toren et al. [29], erosion of materials is caused by local melting and heating of workpiece and tool electrode. Electrode wear is common process, when high MRR with good SQ is required. Khanara et al. [30] have produced ZrB_2–40 wt.% Cu composite tool for machining mild steel, and it has shown a decrease in TWR over pure Cu tool. While comparing with conventional copper tool and ZrB_2–Cu composite tool, average SR and diameter overcut produced on the workpiece are found to be more in the case of composite tool. This work has suggested some results that are promising for the manufacturing industry. Islam et al. [31] have devised a method of removing burrs from drilled holes of carbon fiber–reinforced polymer composites by EDM. They have used four different types of electrodes copper, steel, brass, and aluminum for EDM to remove burrs from the previously drilled holes. Best deburring has been performed by copper electrodes.

Copper electrodes with different geometrical shapes such as round, square, triangular, and diamond with same cross-sectional area of 64 mm^2 have been considered by Khan et al. [32] for studying the effect on the performance of electrode shape on MRR, electrode wear rate (EWR), wear ratio, and SR for mild steel work material. The diamond-shaped tool electrodes with highest EWR and wear ratio are noticed by researchers. Beri et al. [33] have performed EDM experiments on AISI D2 steel with kerosene as dielectric medium using a powder metallurgically produced tool with the composition of 30% copper and 70% tungsten as tool electrode and with the conventional copper electrode. Results have shown that the powder metallurgically produced copper–tungsten tool produced improved performance than the conventional copper tool.

Graphite is becoming common electrode material for die-sinking EDM. It offers greater advantages compared with copper electrodes. Klocke et al. [34] have investigated the specific wear behavior and MRR in detail and related the physical characteristics of the graphite material. Sen et al. [35] have conducted EDM of alloy of Ti–6Al–4V–xB (with $x=0.00$, 0.04, and 0.09) with three different proportions of the alloy as shown using a copper electrode. The analysis has shown that MRR is unaffected by B addition, and it is also found that with B addition, TWR is considerably reduced. For machining CK45 steel, Taweel [36] evaluated the influence of each input parameter on the output response by adopting central composite rotatable design. From the aforementioned analysis, it could be concluded that adequate attention should be given in the selection of tool material for improving the efficiency of EDM process.

1.2.4 MULTISPARK EROSION STUDIES

For each operating pulse, conventional EDM has only one discharge point. Instead of single spark, researchers have proved that by the application of multisparks, EDM efficiency can be improved. Production of complex shaped tool is costly as well as time-consuming, and it requires almost 70% of production cost. In the conventional processes, the operator has to select the tool by trial and error, and this is tedious as well as it may not be accurate. In aerospace and defense industries, parts to be produced are complex and require high accuracy. Here, selection of tool becomes most important.

A novel method of multispark EDM has been first proposed by Kunieda et al. [37] for achieving better MRR and efficient energy consumption. This is achieved by setting up several discharge points for generating pulses. Multiple discharge system with two electrode discharge points in the electric discharge dressing of metal-bonded grinding wheels has been proposed by Suzuki et al. [38]. Using only one pulse generator, for each pulse, two discharges are created at both the gaps between two electrodes and the grinding wheel. Multiple tools are formed by Mohri et al. [39] by dividing the tool electrode into electrically insulated electrodes and connected to pulse generator through resistor by creating multiple discharges simultaneously in each electrode.

Effect of tool wear with different shapes of tools such as triangular, square, rectangular, and circular with size factor consideration with different machining characteristics has been studied by Sohani et al. [40]. Response surface model was used for evaluating the influence of input characteristics on output responses. Higher MRR and lower TWR are obtained for circular-shaped tool, which is considered as the best tool shape. Aerospace alloys such as Inconel 718 and Ti–6Al–4V have been EDM-drilled to find out the influence of single- and multichannel tubular electrodes made of brass and copper materials, and it has been presented by Yilmaz and Okka [41]. A comparative study was performed for analyzing MRR, electrode wear, and microhardness. Investigation has revealed that single-channel electrode has shown good performance in the case of MRR and electrode wear ratio.

Bayramoglu and Duffill [42] have quoted that frame tool can reduce machining time considerably because they have distinct advantages of cost-effectiveness and easy production while using in EDM process. Surface integrity of C-40 steel is studied by Patowari et al. [43] while conducting EDM of the C-40 steel with WC 60%–40 wt.% Cu powder metallurgically prepared composite W-Cu tool, which is used for machining. It is noticed that surface integrity changes due to the deposition of WC and Cu. In order to machine $Al_2O_3/6061Al$ composite, Yan et al. [44] have used rotary disk like electrode and used Taguchi methodology for conducting experiments. Output parameters such as MRR, EWR, Relative Electrode Wear Ratio (REWR), and SR and input parameters such as polarity, peak current, pulse duration, and powder supply voltage are considered for study. Result shows that the performance of output parameters is improved considerably.

An experimental study on EDM of Al 6061 T6 alloy has been carried out by Arooj et al. [45] with cylindrical copper electrode for evaluating the effect of recast layer, MRR, globule diameter, and the interglobule distance. Among four operating parameters such as current, voltage, off time, and on time, it was established that current has greater effect on white layer formation.

1.2.5 SELECTION OF OPTIMIZED PULSE DURATION

In EDM, quality of pulse plays an important role on realization of EDM in a good manner. In conventional machining, machinist has to operate machine in a dexterous manner for stabilizing the machine. For example, sparking pulse is required for material removal mechanism and better surface finish than other pulse types. Usually, EDM gets unstable due to the congregation of debris particles between tool and workpiece gap and gives rough surface finish because of arc and short pulses. EDM process pulses are identified by these controllers, and necessary corrective measures are taken to stabilize the machine.

Electrical pulse condition duration has greater influence on the machining characteristics of EDM process and was first revealed by Son et al. [46]. It was found that pulse duration affected machining performances such as MRR, TWR, and SQ. They have argued that high precise parts could be manufactured by shorter EDM pulse. RC pulse generator is coupled with EDM machine by Gostimirovic et al. [47] for machining manganese–vanadium tool steel workpiece using graphite tool electrode. It has been noted from the observations that highly influencing factors that affect MRR are discharge current and pulse duration.

Ghoreishi and Taber [48] have studied the effects of voltage excitation of the pre-ignition spark pulse on the performance measures such as MRR, EWR, and average SR in EDM process. Researchers have argued that by applying proper voltage excitation, an effective pulse can be produced, and it will increase material erosion and SQ. Possibility of machining with ultralow discharge and workpiece agitation provides an RC discharge circuit, and it has been first investigated by Egashira et al. [49] by maintaining capacitance equal to the machine's stray capacitance, which is equal to 30 pF. A 15-mm-diameter tungsten electrode has been used for machining copper workpiece. Machined surfaces at this condition have produced smooth and crater-free surface. Fonda et al. [50] have conducted experiments to evaluate electrical and thermal properties of Ti–6Al–4V composite by EDM. Specially designed thermocouple has been used for measuring temperature of workpieces at different stages of experiments. It is noted that there is minute change in workpiece temperature during machining. But change in duty factor percentage has resulted change in internal temperature of workpiece.

For achieving good surface finish in EDM process, Muthuramalingam et al. [51] have conducted experiments. Pulse discrimination methods in EDM do not perform well with rotating electrode because of rare occurrence of arcs during the EDM process. This disadvantage has been overcome by Tee et al. [52] by devising a new mechanism for differentiating various types of pulses in the presence of a rotating electrode. Results have shown that this new system gives superior performance in differentiating normal pulses from harmful arcs, open circuit and short circuit pulses, compared with the conventional methods. Shanker et al. [53] have theoretically calculated the shapes of arc at different cross sections in EDM process. It is found that arc produced in EDM process is noncylindrical. They have also created mathematical equations for calculating heat absorbed by cathode and dielectric, and they have also compared electrode wear by theoretical and experimental methods. For studying and evaluating the impotence of pulse wave forms, Bozkurt et al. [54] have created a simulation model for single and multisparks. This has resulted in the study of melting and evaporation electrodes, and thermally protective layers have reduced tool wear.

1.2.6 Vibratory Tool and Workpiece

Workpiece, tool, or dielectric medium are allowed to vibrate ultrasonically, and machining operation done in this condition is called ultrasonic-assisted EDM. In ultrasonic-assisted EDM, higher efficiency in machining operation is achieved by obtaining improvement in dielectric circulation and higher debits removal from work area. For simplification of the equipment and reduction of cost, Zhang et al. [55] have used a DC power supply to produce supply spark with ultrasonic frequency. Normal pulse power is replaced in this mechanism, and it has shown agreeable results. In this experiment, relative motion between the tool and workpiece produces pulse discharge. Krishna and Patowari [56] have prepared tungsten and copper powder metullargically created electrode for giving a coating on mild steel workpiece. The outputs measured are material transfer rate and tool erosion rate. With various input factors, Taguchi method was used for conducting experiments. It is observed that good surface modification has been obtained by this method. Shabgard et al. [57] have compared ultrasonic vibration of workpiece with the conventional EDM process for machining FW4 welding metal. For normal operating characteristics, ultrasonic-assisted EDM (US-EDM) functions better.

With three different types of electrodes such as normal copper electrode, cryogenically cooled copper electrode, and ultrasonic-assisted cryogenically cooled copper electrode, Srivastava et al. [58] have compared EDM performance on M2 grade high-speed steel. Observation by researchers is that UACEDM process gives better SQ. Singh and Singhal [59] have used response surface methodology for optimizing process parameters of rotary ultrasonic EDM for machining alumina ceramic. The

Machining of Engineering Materials

results observed indicate considerable increase in the performance of EDM. Tabar et al. [60] have introduced a new method of applying ultrasonic vibration on tool and workpiece for improving material erosion rate. Researchers have shown that applying ultrasonic vibration to both tool and workpiece causes collapse of air bubble in a smooth manner. This expels debris from the machined area, and it accelerates material removal.

1.2.7 INTRODUCING SERVO CONTROL MECHANISM

For achieving stability in EDM by controlling the discharge gap and quality of spark, usually servo control mechanism is used in machining process. Maintaining constant gap between the electrodes can be achieved by an adaptive control system mechanism that maintains a desired spark gap and prevents harmful arcing and short circuit. Undesirable machining conditions can be avoided by incorporating a self-tuning adaptive servo control system and a better flushing model. An optimized adoptive control system will detect unwanted signals in advance and take necessary preventive actions before undesirable machining conditions occur. With the help of EDM oil (dielectric) pressure, Hayakawa et al. [61] have proposed a new online gap measurement and control model for minimizing short-circuiting and debris flashing problems. This has been proved to be an efficient method in ED machining. Maximum variance control to minimum-variance, pole-placement coupled control, and two-step prediction control have been used for stabilizing the EDM process by Zhou et al. [62]. Adaptive control systems have the capacity of correcting sparking process by minimum variance control, and superior machining ability is achieved by this method. This process has been proposed by Rajurkar [63]. While producing small holes and medium-sized cavities, higher productivity can be achieved compared with model reference adaptive controller. This control system can be accommodated in any EDM unit. Several other researchers have also contributed on servo control mechanism for improving machining performance.

In order to replace the conventional solid electrodes, electrodes formed by copper wire bunch are used in experiments, and it has been proved to be more economical than the conventional solid electrodes by Dursun and Cogun [64]. It is found that a large number of wires in electrodes increase the machining area, thereby reducing rough machining process, machining cost, and time. Electrodes required in machining process and the delay in starting the process are reduced considerably by selecting wire bunch electrodes. Manufacturing cost involved in this wire type of electrode is lower compared with the conventional solid electrodes.

1.2.8 MAGNETIC FIELD–BASED ELECTRICAL DISCHARGE MACHINING

Removal of debris in EDM process is a Himalayan task for researchers. Abnormal electrical discharges may occur, due to the accumulation of debris within the work area and in the dielectric fluid. This process actually reduces the machining gap. This would create instability in EDM process. For achieving high-quality machining and precise as well as efficient managing of debris in the machining gap, dielectric fluids should be forced out of machining area effectively and completely. Introduction of magnetic force for cleaning the machining area in EDM process is a new area of research.

Peak current and pulse duration are selected as input parameters for analyzing their effects on output parameters under the influence of magnetic field coupled with the EDM for machining Al-based metal–matrix composites, and this has been analyzed by Bains et al. [65]. Reduction in recast layer formation with improved MRR is the result, when higher spark energy under magnetic field environment is applied. EDM under electromagnetic field for machining Ti–6Al–4V workpiece with copper tool for smooth expulsion of debris from machining area has been experimented by Naidu et al. [66]. A model equation has been constructed to get relation between MRR and SR with the input parameters. Researchers have found that ionizing plasma channel and smooth expulsion of debris from the adjacent areas of machining would definitely improve the process.

Experimental trials have been divided into low-energy regime, middle-energy regime, and high-energy regime by Teimouri and Baseri [67] for studying the combined effects of tool rotation and external magnetic field on EDM performance. Experiments have proved that output parameters have improved considerably. In order to make machining fast and for getting good surface finish to the component, Bhatt et al. [68] have proposed a method of powder mixed dielectric with externally assisted magnetic field EDM for machining complex shapes. They have successfully machined various die steels such as AISI D2, AISI D3, and H13 using different metal powders in dielectric such as graphite, tungsten (W), and titanium (Ti). This method has improved the performance of EDM considerably. Govindan et al. [69] have conducted experiments in EDM with single-spark and with single-pulse generator. The experiments have been conducted in dry condition by supplying oxygen gas to the electrode area, and the experiments are repeated in wet condition by applying liquid dielectric surrounding the spark column. In all the experiments, a pulsating magnetic field is applied. They have selected stainless steel 304 as workpiece and electrolyte copper as tool. Considerable increase in performance has been achieved in dry EDM magnetic field–assisted condition.

1.2.9 SPECIAL TOOLS

Machining industrial parts that are made from glass, plastics, ceramics, composite materials, and nonferrous metals in large scale is a new research area. A two-stage EDM process consisting rough cutting by wire electrode EDM and then subsequent finishing operations using rotational disc electrode EDM has been introduced by Mamalis et al. [70] for machining diamond polycrystalline superhard material. This method can be employed for machining superhard materials. By using copper and graphite as electrodes as well as using planetary tool actuation in EDM, square-shaped cavities have been generated on Ti6Al4V alloy, and this research has been conducted by Mathai et al. [71]. Though graphite electrodes are found to be more preferable with high MRR, it is also noted that copper electrodes also have some feasibility. Song et al. [72] devised a novel method for turning of stainless steel 304 using a brass strip electrode. Deionized water is the dielectric used in the EDM

Machining of Engineering Materials

process. It is clear that this method is more suitable than the conventional wire EDM process, and it is less expensive.

1.2.10 CNC-Controlled Electrical Discharge Machining

Using open CNC and soft CNC technology, Huang et al. [73] have developed software for multiaxis EDM with CNC systems for machining of rimmed turbine blisks. Linux and RT-Linux are used for the design of platforms and incorporated into five-axis EDM CNC machine for machining of rimmed turbine blisks. Researchers have found that this software is reliable and stable and can be used at any condition.

Development of CNC EDM for machining-complicated shapes by developing suitable paths has been developed by Ding and Jiang [74] with Unigraphis software. This software can be used for the generation of suitable tool paths, adjustment of pulse, and discharge length, thereby increasing productivity and reduction of cost.

Wang et al. [75] have shown that machining accuracy depends on the stability of an EDM machine, and this, in turn, is influenced by static and dynamic characteristics. SolidWorks is used to build the three-dimensional model with the help of ANSYS workbench finite element method, and it has been employed to analyze the static and dynamic characteristics of the EDM machine. For optimizing the performance of EDM machines, this method is found to be useful. Ding et al. [76] have introduced a new algorithm to detect sharp corner for manufacturing tool electrode for injection molds using EDM process. Tool actually consists of electrode and tool holder, and it is manufactured as one piece. Using this algorithm, 3D image of tool can be visualized and can be corrected according to the requirements. Mild steel workpieces and low-cost "frame-type" copper tools have been used by Bayramoglu et al. [77] in lower powered CNC EDM. The experiment shows that in the case of production of cavities, an improved MRR and a better surface finish are obtained in this research.

1.2.11 Selection of Dielectric Medium

There are many functions in EDM process where dielectric fluid can serve and the one that acts as a medium for controlling sparks. It acts as a cooling medium to remove excess heat from the molten metal debris, and that is generated because of repeated sparks. Dielectric fluid removes away the debris from work area and makes the operations smoother. Mineral oil and kerosene are two dielectric fluids commonly used in EDM process. They have low flash point and high volatility and are nonhazardous. Researchers have found that addition of metal or alloy powder to dielectric will improve the efficiency of EDM process. Metal powder, which is added in dielectric, fills the gap between the tool and electrode. When voltage is applied, it gets energized and behaves in a zigzag manner by increasing of discharge gap. Long et al. [78] have shown that titanium powder mixed in dielectric increases MRR and reduces TWR. Influence of TiO_2 nanopowder in dielectric fluid coupled with tool rotation, and its effects on MRR, TWR, and SR are investigated by Baseri et al. [79]. The addition of aluminum powder in kerosene dielectric and EDM of aluminum/alumina metal–matrix composite are the research area of Talla et al. [80].

It is observed considerable improvement in MRR and SR compared with that of conventional EDM. Discharge current, surfactant, and graphite powder concentration in dielectric medium are used as input process parameters for EDM of Ti–6Al–4V, and it has been studied by Kolli and Kumar [81].

Graphite powder–added kerosene is used as dielectric by Unses and Cogun [82] for conducting EDM experiments for improving the SQs of Ti–6Al–4V alloy. This suspended dielectric during machining has shown considerable increase in MRR and surface finish and decrease in relative wear. Nihal Ekmekci and Bülent Ekmekci [83] have performed EDM of Ti–6Al–4V work material using hydroxyapatite (HA) powder suspension in deionized water as dielectric liquid. The investigations have revealed that there is transfer of powder particles from dielectric liquid. Very thin recast layer around 1–2 mm in thickness has been achieved by adding surfactant with Al powder by Wu et al. [84] in EDM. Effect of distribution electrical discharging energy is more visible in this experiment.

Graphite powder as additive in dielectric and its effect on improving the surface properties of superalloy Super Co 605 in EDM have been investigated by Anoop Singh et al. [85]. Zhang et al. [86] have used thin-walled pipe electrode with internal hole in ultrasonic-assisted EDM in order to improve MRR. In this experiment, gas is used as dielectric and is applied through internal hole. Emission of harmful vapor such as CO and CH_4 will occur as a result of decomposition of hydrocarbon oils such as kerosene. This can be overcome by replacing hydrocarbon oil with water. Researchers have compared performances of kerosene and distilled water as dielectric in EDM. Jeswani [87] has observed that a higher MRR and a lower wear ratio than in kerosene is obtained when high pulse energy range is used. When distilled water is used as dielectric, poor machining accuracy and better surface finish are observed. For analyzing the performance of water as dielectric in EDM, Guo et al. [88] have conducted machining of Ti–35V–15Cr alloy with graphite as tool electrode. They analyzed performance of working fluid based on thermal conductivity and bubble formation during the process. It is proved water as an alternative to conventional working fluids.

Using kerosene as dielectric would cause severe problems such as pollution, environmental damage, and fire hazard. In order to avoid these phenomena, Lin et al. [89] applied compressed air, oxygen, and argon as gas dielectric that are passed to the machining zone using a designed system for machining SKD 61 steel. In order to achieve initial SQ and parallelism, workpiece surface is ground in a perfect manner. EDM assisted with ultrasonic vibration and using cylindrical electrolytic copper as tool electrode have proved to be an efficient machining system. Zhang et al. [90] have prepared (water in oil) W/O emulsions for sinking EDM and compared machining characteristics with conventional kerosene.

Effect of cooling copper electrode with cold nitrogen air–mixed gas on EDM of W6Mo5Cr4V2 high-speed steel has been studied by Lil et al. [91]. Result indicates that cooling effect could considerably influence MRR, EW, and EWR. Cracks in machined surface of the workpiece are observed at a temperature below −80°C. Cryogenic cooling of electrode could reduce electrode wear more than 25%, and it is observed by Abdulkareem et al. [92], while studying the cooling effect of copper electrode on ED machining of titanium alloy (Ti–6Al–4V). Ng et al. [93] have used Canola BD and sunflower BD as dielectric fluid.

1.3 APPLICATION OF ELECTRICAL DISCHARGE MACHINING FOR BIOMATERIALS

EDM finds good application in machining biomedical components. Biomedical components are usually very hard and difficult to machine by conventional methods. Complicated shapes and very precise accuracy are required in transplanting bone-related plates to human body. EDM has good scope in this area. Pradhan et al. [94] have investigated dry machining of Ti–6Al–4V alloy using SNMA120408 grade inserts for biomedical applications. Basak et al. [95] investigated machining of titanium alloy with wire electrical discharge machining. By varying the flushing pressure, wire tension, and pulse on time, they have studied the effects of kerf width, MRR, and discharge gap. It is found deformed wire after experiment and recast later with cracks and holes. Prakesh et al. [96] have used powder-mixed EDM for studying its effect on a b-Ti-based implant for producing a biomimetic nanoporous bioceramic surface. It has been successfully proved by them that a natural bonelike surface structure can be produced by this method. Aliyu et al. [97] have successfully studied machining of bulk metallic glasses with HA-mixed EDM (HA-EDM). The surface produced by HA-EDM is expected to facilitate higher tissue ingrowth and bone-implant adhesion.

Prakesh et al. [98] have fabricated an alloy by depositing a nanohydroxyapatite (nHA) coating on biodegradable Mg–Zn–Mn alloy. The main purpose of coating is to reduce the corrosion of implants. It is proved that such coating helps in reducing corrosion at a considerable level. Prakesh et al. [99] have used powder-mixed EDM and nondominated sorting genetic algorithm for fabricating biocompatible surface on β-phase Ti alloy. After optimizing and confirmation, the result shows 184% increase in microhardness. They have used response surface methodology for predicting the results. EDM, PMED, and W-EDM found potential to enhance the Mg- and Ti-based implants for biomedical applications [96,99,100–126].

1.4 CONCLUSIONS

EDM plays a significant role in precision industry, as it can create mirror image of tool on the workpiece so that it can reproduce complicated shapes. EDM is generally employed for machining very hard and brittle workpieces that are difficult to machine by traditional machining techniques. Even though the main disadvantage is that it can machine only conductive materials, researchers are now going on for machining nonconductive ceramic materials such as alumina and zirconia. Limitations of EDM process include creation of surface cracks, metallurgical changes on surface and subsurface regions, heat-affected zone, recast layer on machined surface, residual stresses, low MRR, high tool wear, and high SR. Even though a noncontact process, series of sparks causes stresses in the workpiece, this causes residual stresses in the workpiece. Slow MRR, high tool wear, poor surface quality, formation of craters, and recast layer are the major disadvantages of the process. Researchers have solved these problems by introducing dry EDM, near-dry EDM, ultrasonic-assisted EDM, and powder-mixed EDM. This chapter provides a close review of literature developments in past and present decades. The present trend is automation of EDM process by CNC machines, and new areas of ceramic materials have evolved. A new research area in biomedical engineering for machining human implants has also been discussed in detail.

REFERENCES

1. S. Suresh Kumar, M. Uthayakumar, S. Thirumalai Kumaran and P. Parameswaran. Electrical discharge machining of Al (6351)–SiC–B$_4$C hybrid composite. *Materials and Manufacturing Processes*, 2014, 29 (11–12), 1395–1400.
2. K. Ojha, R.K. Garg and K.K. Singh. MRR improvement in sinking electrical discharge machining: a review. *Journal of Minerals & Materials Characterization & Engineering*, 2010, 9 (8), 709–739.
3. A. Pandey and S. Singh. Current research trends in variants of electrical discharge machining: a review. *International Journal of Engineering Science and Technology*, 2010, 2 (6), 2172–2191.
4. J.S. Soni and G. Chakraverti. Experimental investigation on migration of material during EDM of die steel (T215 Cr12). *Journal of Materials Processing Technology*, 1996, 56, 439–451.
5. J. Marafona and C. Wykes. A new method of optimizing material removal rate using EDM with copper tungsten electrodes. *International Journal of Machine Tools & Manufacture*, 2000, 40 (2), 153–164.
6. J. Marafona. Black layer characterization and electrode wear ratio in electrical discharge machining (EDM). *Journal of Materials Processing Technology*, 2007, 184 (1–3), 27–31.
7. N. Mohri, M. Suzuki, M. Furuya and A. Kobayashi. Electrode wear process in electrical discharge machining. *CIRP Annals - Manufacturing Technology*, 1995, 44 (1), 165–168.
8. N. Mohri, Saito, T. Takawashi and K. Kobayashi. Mirror-like finishing by EDM. *Proceedings of the Twenty-Fifth International Machine Tool Design and Research Conference*, 1985 (pp. 329–336). Palgrave, London.
9. F.L. Amorim, V.A. Dalcin, P. Soares and A. Luciano. Mendes, Surface modification of tool steel by electrical discharge machining with molybdenum powder mixed in dielectric fluid. *The International Journal of Advanced Manufacturing Technology*, 2017, 91, 341–350.
10. B. Jha, K. Ram and M. Rao. An over view of technology and research in electrode design and manufacturing in sinking electrical discharge machining. *Journal of Engineering Science and Technology Review*, 2011, 4 (2), 118–130.
11. S. Vinoth Kumar and M. Pradeep Kumar. Optimization of cryogenic cooled EDM process parameters using grey relational analysis. *Journal of Mechanical Science and Technology*, September 2014, 28 (9), 3777–3784.
12. A. Majumder. Process parameter optimization during EDM of AISI 316 LN stainless steel by using fuzzy based multi-objective PSO. *Journal of Mechanical Science and Technology*, 2013, 27 (7), 2143–2151.
13. R. Teimouri and H. Baseri. Optimization of magnetic field assisted EDM using the continuous ACO algorithm. *Applied Soft Computing*, 2014, 14, 381–389.
14. M.R. Shabgard, B. Sadizadeh and H. Kakoulvand. The effect of ultrasonic vibration of workpiece in electrical discharge machining of AISIH13 tool steel. *World Academy of Science, Engineering and Technology*, 2009, 3, 332–336.
15. J.T. Huang, Y.S. Liao and W.J. Hsue. Determination of finish-cutting operation number and machining parameters setting in wire electrical discharge machining. *Journal of Materials Processing Technology*, 1999, 87, 69–81.

Machining of Engineering Materials

16. V.R. Khullar, N. Sharma, S. Kishore and R. Sharma. RSM- and NSGA-II-based multiple performance characteristics optimization of EDM parameters for AISI 5160. *Arabian Journal for Science and Engineering*, May 2017, 42 (5), 1917–1928.
17. Jagadish and A. Ray. Optimization of process parameters of green electrical discharge machining using principal component analysis (PCA). *The International Journal of Advanced Manufacturing Technology*, 2016, 87 (5), 1299–1311.
18. X.-P. Dang. Constrained multi-objective optimization of EDM process parameters using kriging model and particle swarm algorithm. *Materials and Manufacturing Processes*, 2018, 33 (4), 397–404.
19. D.K. Panda. Modelling and optimization of multiple process attributes of electro discharge machining process by using a new hybrid approach of neuro–grey modeling. *Materials and Manufacturing Processes*, 2010, 25 (6), 450–461.
20. S. Tripathy and D.K. Tripathy. Multi-response optimization of machining process parameters for powder mixed electro-discharge machining of H-11 die steel using grey relational analysis and topsis. *Machining Science and Technology*, 2017, 21 (3), 362–384.
21. R. Manivannan and M. Pradeep Kumar. Multi-attribute decision making of cryogenically cooled Micro-EDM drilling process parameters using TOPSIS method. *Materials and Manufacturing Processes*, 2017, 32 (2), 209–215.
22. E. Aliakbari, H. Baseri. Optimization of machining parameters in rotary EDM process by using the Taguchi method. *The International Journal of Advanced Manufacturing Technology*, October 2012, 62 (9–12), 1041–1053.
23. J.D. Marafona and A. Araújo. Influence of work piece hardness on EDM performance. *International Journal of Machine Tools and Manufacture*, July 2009, 49 (9), 744–748.
24. K. Al-Ghamdi and O. Taylan. A comparative study on modelling material removal rate by ANFIS and polynomial methods in electrical discharge machining process. *Computers & Industrial Engineering*, 2015, 79, 27–41.
25. S. Gopalakannan and T. Senthilvelan. Effect of electrode materials on electric discharge machining of 316L and 17-4 PH stainless steels. *Journal of Minerals and Materials Characterization and Engineering*, 2012, 11, 685–690.
26. A. Torres, I. Puertas and C.J. Luis. EDM machinability and surface roughness analysis of INCONEL 600 using graphite electrodes. *The International Journal of Advanced Manufacturing Technology*, June 2016, 84 (9), 2671–2688.
27. N. Pragadish and M. Pradeep Kumar. Surface characteristics analysis of dry EDM AISI D2 steel using modified tool design. *Journal of Mechanical Science and Technology*, April 2015, 29 (4), 1737–1743.
28. N.K. Singh, P.M. Pandey and K.K. Singh. Experimental investigations into the performance of EDM using argon gas-assisted perforated electrodes. *Materials and Manufacturing Processes*, 2017, 32 (9), 940–951.
29. M. Toren, Y. Zvirin and Y. Winograd. Melting and evaporation phenomena during electrical erosion. *Transactions of ASME Journal of Heat Transfer*, 1975, 97, 576–581.
30. A.K. Khanara, B.R. Sarkar, B. Bhattacharya and L.C. Pathak. Performance of ZrB2-Cu composite as an EDM tool. *Journal of Material Processing Technology*, 2007, 183, 122–126.
31. M.M. Islam, C.P. Li, S.J. Won and T.J. Ko. A deburring strategy in drilled hole of CFRP composites using EDM process. *Journal of Alloys and Compounds*, May 2017, 5, 477–485.
32. A.A. Khan, M.Y. Ali and M.M. Haque. A study of electrode shape configuration on the performance of die sinking EDM. *International Journal of Mechanical and Materials Engineering*, 2009, 4 (1), 19–23.
33. N. Beri, S. Maheshwari, C. Sharma and A. Kumar. Performance evaluation of powder metallurgy electrode in electrical discharge machining of AISI D2 steel using Taguchi method. *International Journal of Mechanical, Aerospace, Industrial, Mechatronic and Manufacturing Engineering*, 2008, 2 (2), 225–229.

34. F. Klocke, M. Schwade, A. Klinka and D. Veselovaca. Analysis of material removal rate and electrode wear in sinking EDM roughing strategies using different graphite grades. *Procedia CIRP*, 2013, 6, 163–167.

35. I. Sen, G. Karthikeyan, J. Ramkumar and R. Balasubramaniam. A study on machinability of B-modified Ti-6Al-4V alloys by EDM. *Materials and Manufacturing Processes*, 2012, 27 (3), 348–354.

36. T.A. El-Taweel. Multi-response optimization of EDM with Al–Cu–Si–TiC P/M composite electrode. *The International Journal of Advanced Manufacturing Technology*, September 2009, 44 (1–2), 100–113.

37. M. Kunieda and H. Muto. Development of multi spark EDM. *CIRP Annals*, 2000, 49 (1), 119–122.

38. K. Suzuki, N. Mohri, T. Uematsu and T. Nakagawa. ED truing method with twin electrodes. *Preprint of Autumn Meeting of JSPE (Japanese)*, 1985, 575–578.

39. N. Mohri, H. Takezawa, K. Furutani, Y. Ito and T. Sata. New process of additive and removal machining by EDM with a thin electrode. *CIRP Annals Manufacturing Technology*, 2000, 49 (1), 123–126.

40. M.S. Sohani, V.N. Gaitonde, B. Siddeswarappa and A.S. Deshpande. Investigations into the effect of tool shapes with size factor consideration in sink electrical discharge machining (EDM) process. *International Journal Advanced Manufacturing Technology*, 2009, 45, 1131–1145.

41. O. Yilmaz and M.A. Okka. Effect of single and multi-channel electrodes application on EDM fast hole drilling performance. *International Journal Advanced Manufacturing Technology*, 2010, 51,185–194.

42. M. Bayramoglu and A.W. Duffill. Systematic investigation on the use of cylindrical tools for the production of 3D complex shapes on CNC EDM machines. *International Journal of Machine Tools Manufacture*, 1994, 34 (3), 327–339.

43. P.K. Patowari, P. Saha and P.K. Mishra. An experimental investigation of surface modification of C-40 steel using W–Cu powder metallurgy sintered compact tools in EDM. *The International Journal of Advanced Manufacturing Technology*, September 2015, 80 (1–4), 343–360.

44. B.H. Yan, C.C. Wang, W.D. Liu and F.Y. Huang. Machining characteristics of Al2O3/6061Al composite using rotary EDM with a disklike electrode. *The International Journal of Advanced Manufacturing Technology*, 2000, 16 (5), 322–333.

45. S. Arooj, M. Shah, S. Sadiq, S.H.I. Jaffery and S. Khushnood. Effect of current in the EDM machining of aluminum 6061 T6 and its effect on the surface morphology. *Arabian Journal for Science and Engineering*, 2014, 39, 4187–4199.

46. S.M. Son, H.S. Lim, A.S. Kumar and M. Rahman. Influences of pulsed power condition on the machining properties in micro EDM. *Journal of Materials Processing Technology*, 2007, 190, 73–76.

47. M. Gostimirovic, P. Kovac, M. Sekulic and B. Skoric. Influence of discharge energy on machining characteristics in EDM. *Journal of Mechanical Science and Technology*, January 2012, 26 (1), 173–179.

48. M. Ghoreishi and C. Tabari. Investigation into the effect of voltage excitation of pre-ignition spark pulse on the electro-discharge machining (EDM) process. *Materials and Manufacturing Processes*, 2007, 22 (7–8), 833–841.

49. K. Egashira, A. Matsugasako, H. Tsuchiya and M. Miyazaki. Electrical discharge machining with ultralow discharge energy. *Precision Engineering*, 2006, 30, 414–420.

50. P. Fonda, Z. Wang, K. Yamazaki and Y. Akutsu. A fundamental study on Ti–6Al–4V's thermal and electrical properties and their relation to EDM productivity. *Journal of Materials Processing Technology*, 2008, 202, 583–589.

Machining of Engineering Materials

51. T. Muthuramalingam and B. Mohan. Performance analysis of ISO current pulse generator on machining characteristics in EDM process. *Archives of Civil and Mechanical Engineering*, 2004, 14, 383–390.

52. K.T. Pey Tee, R. Hosseinnezhad, M. Brandt and J. Mo. Pulse discrimination for electrical discharge machining with rotating electrode. *Machining Science and Technology*, 2013, 17 (2), 292–311.

53. P. Shankar, V.K. Jain and T. Sundararajan. Analysis of spark profiles during EDM process. *Machining Science and Technology*, 1997, 1 (2), 195–217.

54. B. Bozkurt, A.M. Gadalla and P.T. Eubank. Simulation of erosions in a single discharge EDM process. *Materials and Manufacturing Processes*, 1996, 11 (4), 555–563.

55. Q.H. Zhang, R. Du, J.H. Zhang and Q. Zhang. An investigation of ultrasonic-assisted electrical discharge machining in gas. *International Journal of Machine Tools & Manufacture*, 2006, 46, 1582–1588.

56. M. Eswara Krishna and P.K. Patowari. Parametric optimization of electric discharge coating process with powder metallurgy tools using Taguchi analysis. *Surface Engineering*, 2013, 29 (9), 703–711.

57. M. Shabgard, H. Kakolvand, M. Seyedzavvarand and R.M. Shotorbani. Ultrasonic assisted EDM: effect of the work piece vibration in the machining characteristics of FW4 Welded Metal. *Frontiers of Mechanical Engineering*, December 2011, 6 (4), 419–428.

58. V. Srivastava and P.M. Pandey. Effect of process parameters on the performance of EDM process with ultrasonic assisted cryogenically cooled electrode. *Materials and Manufacturing Processes*, August 2012, 14 (3), 393–402.

59. R.P. Singh and S. Singhal. Investigation of machining characteristics in rotary ultrasonic machining of alumina ceramic. *Materials and Manufacturing Processes*, 2017, 32 (3), 309–326.

60. M.T. Shervani-Tabar, K. Maghsoudi and M.R. Shabgard. Effects of simultaneous ultrasonic vibration of the tool and the work piece in ultrasonic assisted EDM. *International Journal for Computational Methods in Engineering Science and Mechanics*, 2013, 14 (1), 1–9.

61. S. Hayakawa, M. Takahashi, F. Itoigawa and T. Nakamura. Study on EDM phenomena with in-process measurement of gap distance. *Journal of Materials Processing Technology*, 2004, 149, 250–255.

62. M. Zhou and F. Han. Adaptive control for EDM process with a self-tuning regulator. *International Journal of Machine Tools and Manufacture*, 2009, 49 (6), 462–469.

63. K.P. Rajurkar, W.M. Wang and R.P. Lindsay. A new model reference adaptive control of EDM. *CIRP Annals - Manufacturing Technology*, 1989, 38 (1), 183–186.

64. K. Dursun and C. Cogun. Use of wire bunch electrodes in electric discharge machining. *Rapid Prototyping Journal*, 2009, 15 (4), 291–298.

65. P. Singh Bains, S.S. Sidhu and H.S. Payal. Study of magnetic field-assisted ED machining of metal matrix composites. *Materials and Manufacturing Processes*, 2016, 31 (14), 1889–1894.

66. S. Vignesh, K. Naidu Vipindas, R. Manu and J. Mathew. Experimental study on varying electromagnetic field assisted die sinking EDM. *5th International & 26th All India Manufacturing Technology, Design and Research Conference (AIMTDR 2014)* December 12th–14th, 2014, IIT Guwahati, Assam, India.

67. R. Teimouri and H. Baseri. Experimental study of rotary magnetic field-assisted dry EDM with ultrasonic vibration of work piece. *The International Journal of Advanced Manufacturing Technology*, 2013, 67 (5), 1371–1384.

68. G. Bhatt, A. Batish and A. Bhattacharya. Experimental investigation of magnetic field assisted powder mixed electric discharge machining. *Particulate Science and Technology*, 2015, 33, 246–256.

69. P. Govindan, A. Gupta, S.S. Joshi, A. Malshe and K.P. Rajurkar. Single-spark analysis of removal phenomenon in magnetic field assisted dry EDM. *Journal of Material Processing Technology*, July 2013, 213 (7), 1048–1058.
70. A.G. Mamalis, A.I. Grabchenko, M.G. Magazeev, N.V. Krukova, J. Prohàszká and N.M. Vaxevanidis. Two-stage electro-discharge machining fabricating super hard cutting tools. *Journal of Materials Processing Technology*, 2004, 14, 318–325.
71. V.J. Mathai, H.K. Dave and K.P. Desai. Experimental investigations on EDM of Ti6Al4V with planetary tool actuation. *Journal of the Brazilian Society of Mechanical Sciences and Engineering*, 2017, 39 (9), 3467–3490.
72. K.Y. Song, D.K. Chung, M.S. Park and C.N. Chu. EDM turning using a strip electrode. *Journal of Materials Processing Technology*, 2013, 213 (9), 1495–1500.
73. H. Huang, G. Chi and Z. Wang. Development and application of software for open and soft multi-axis EDM CNC systems. *The International Journal of Advanced Manufacturing Technology*, 2016, 86 (9–12), 2689–2700.
74. S. Ding and R. Jiang. Tool path generation for 4-axis contour EDM rough machining. *International Journal of Machine Tools and Manufacture*, 2004, 44 (14), 1493–1502.
75. J.H. Wang, G. Liz, F. Liu, Y.S. Zhao, Z.J. Cheng, J.Y. Liu and Y. Li. Studies of static and dynamic characteristics of the EDM machine based on the ANSYS workbench. *Strength of Materials*, January 2015, 47 (1), 87–93.
76. X.M. Ding, J.Y.H. Fuh, K.S. Lee, Y.F. Zhang and A.Y.C. Nee. A computer-aided EDM electrode design system for mold manufacturing. *International Journal of Production Research*, 2000, 38 (13), 3079–3092.
77. M. Bayramoglu and W. Duffill. Manufacturing linear and circular contours using CNC EDM and frame type tools. *International Journal of Machine Tools Manufacture*, 1995, 35 (8), 1125–1136.
78. B.T. Long, N.H. Phan, N. Cuong and N.D. Toan. Surface quality analysis of die steels in powder-mixed electrical discharge machining using titan powder in fine machining. *Advances in Mechanical Engineering*, 2016, 8 (7), 1–13.
79. H. Baseri and S. Sadeghian. Effects of nano powder TiO_2-mixed dielectric and rotary tool on EDM. *The International Journal of Advanced Manufacturing Technology*, March 2016, 83 (1–4), 519–528.
80. G. Talla, D.K. Sahoo, S. Gangopadhyay and C.K. Biswas. Modeling and multi-objective optimization of powder mixed electric discharge machining process of aluminum/alumina metal matrix composite. *Engineering Science and Technology, an International Journal*, 2015, 18, 369–373.
81. M. Kolli and A. Kumar. Effect of dielectric fluid with surfactant and graphite powder on electrical discharge machining of titanium alloy using Taguchi method. *Engineering Science and Technology, an International Journal*, 2015, 18, 524–535.
82. E. Unses and C. Cogun. Improvement of electric discharge machining (EDM) performance of Ti-6Al-4V alloy with added graphite powder to dielectric. *Strojniški vestnik - Journal of Mechanical Engineering*, 2015, 61, 409–418.
83. N. Ekmekci and B. Ekmekci. Electrical discharge machining of Ti6Al4V in hydroxyapatite powder mixed dielectric liquid. *Materials and Manufacturing Processes*, 2016, 31 (13), 1663–1670.
84. K.L. Wu, B.H. Yan, F.Y. Huang and S.C. Chen. Improvement of surface finish on SKD steel using electro-discharge machining with aluminum and surfactant added dielectric. *International Journal of Machine Tools and Manufacture*, 2005, 45, 1195–1201.
85. A.K. Singh, S. Kumar and V.P. Singh. Optimization of parameters using conductive powder in dielectric for EDM of super Co 605 with multiple quality characteristics. *Materials and Manufacturing Processes*, 2014, 29 (3), 267–273.

86. Y. Zhang, Y. Liu, Y. Shen, R. Ji, Z. Li and C. Zheng. Investigation on the influence of the dielectrics on the material removal characteristics of EDM. *Journal of Materials Processing Technology*, 2014, 214, 1052–1061.
87. M.L. Jeswani. Electrical discharge machining in distilled water. *Wear*, 1981, 72, 81–88.
88. C. Guo, S. Di and D. Wei. Study of electrical discharge machining performance in water-based working fluid. *Materials and Manufacturing Processes*, 2016, 31 (14), 1865–1871.
89. Y.-C. Lin, J.-C. Hung, H.-M. Lee, A.-C. Wang and J.-T. Chen. Machining characteristics of a hybrid process of EDM in gas combined with ultrasonic vibration. *The International Journal of Advanced Manufacturing Technology*, 2017, 92, 2801–2808. DOI: 10.1007/s00170-017-0369-z.
90. Y. Zhang, Y. Liu, R. Ji, B. Cai and Y. Shen. Sinking EDM in water-in-oil emulsion. The *International Journal of Advanced Manufacturing Technology*, March 2013, 65 (5), 705–716.
91. L. Li, F.Q. Bi, L. Feng and X. Bai. Effect of low-temperature gas-cooled electrode on the performance of EDM process. *The International Journal of Advanced Manufacturing Technology*, September 2016, 86 (1), 717–722.
92. S. Abdulkareem, A.A Khan and M. Konneh. Reducing electrode wear ratio using cryogenic cooling during electrical discharge machining. *International Journal of Advanced Manufacturing Technology*, 2009, 45, 1146–1151.
93. P.S. Ng, S.A. Kong and S.H. Yeo. Investigation of biodiesel dielectric in sustainable electrical discharge machining. *The International Journal of Advanced Manufacturing Technology*, June 2017, 90 (9–12), 2549–2556.
94. S. Pradhan, S. Singh, C. Prakash, G. Królczyk, A. Pramanik and C.I. Pruncu. Investigation of machining characteristics of hard-to-machine Ti-6Al-4V-ELI alloy for biomedical applications. *Journal of Material Research and Technology*, 2019, 8 (5), 4849–4862.
95. A. Basak, A. Pramanik and C. Prakash. Surface, kerf width and material removal rate of Ti6Al4V titanium alloy generated by wire electrical discharge machining. *Heliyon*, 2019, 5.
96. C. Prakash, H.K. Kansal, B.S. Pabla and S. Puri. Processing and characterization of novel biomimetic nanoporous bioceramic surface on β-Ti implant by powder mixed electric discharge machining. *Journal of Materials Engineering and Performance*, 2015, 24 (9), 3622–3633.
97. A.A. Aliyu, A.M. Abdul-Rani, T.L. Ginta, C. Prakash, E. Axinte and R. Fua-Nizan. Fabrication of nanoporosities on metallic glass surface by hydroxyapatite mixed EDM for orthopedic application. *International Medical Device and Technology Conference*, 2017, 168–171.
98. C. Prakash, S. Singh, B.S. Pabla and M.S. Uddin. Synthesis, characterization, corrosion and bioactivity investigation of nano-HA coating deposited on biodegradable Mg-Zn-Mn alloy. *Surface and Coatings Technology*, 2018, 346, 9–18.
99. C. Prakash, S. Singh, M. Singh, K. Verma, B. Chaudhary and S. Singh. Multi-objective particle swarm optimization of EDM parameters to deposit HA-coating on biodegradable Mg-alloy. *Vacuum*, 2018, 158, 180–190.
100. C. Prakash, H.K. Kansal, B.S. Pabla and S. Puri. Multi-objective optimization of powder mixed electric discharge machining parameters for fabrication of biocompatible layer on β-Ti alloy using NSGA-II coupled with Taguchi based response surface methodology. *Journal of Mechanical Science and Technology*, 2016, 30 (9), 4195–4204. DOI: 10.1007/s12206-016-0831-0.
101. C. Prakash and M.S. Uddin. Surface modification of β-phase Ti implant by hydroxyapatite mixed electric discharge machining to enhance the corrosion resistance and in-vitro bioactivity. *Surface Coating and Technology*, 2017, 236 (Part A), 134–145.

102. C., Prakash, S. Singh, R. Singh, S. Ramakrishna, B.S. Pabla, S. Puri and M.S. Uddin. *Biomanufacturing*. 2019. Springer International Publishing, Switzerland.
103. Singh H., Singh S., Prakash C. (2019) Current Trends in Biomaterials and Biomanufacturing. In: Prakash C. et al. (eds) *Biomanufacturing*. Springer, Cham.
104. C. Prakash, H.K. Kansal, B.S. Pabla, S. Puri and A. Aggarwal. Electric discharge machining–a potential choice for surface modification of metallic implants for orthopedic applications: a review. *Proceedings of the Institution of Mechanical Engineers, Part B: Journal of Engineering Manufacture*, 2016, 230 (2), 331–353.
105. C. Prakash, H.K. Kansal, B.S. Pabla and S. Puri. Experimental investigations in powder mixed electric discharge machining of Ti–35Nb–7Ta–5Zrβ-titanium alloy. *Materials and Manufacturing Processes*, 2017, 32 (3), 274–285.
106. C. Prakash, H.K. Kansal, B.S. Pabla and S. Puri. Powder mixed electric discharge machining: an innovative surface modification technique to enhance fatigue performance and bioactivity of β-Ti implant for orthopedics application. *Journal of Computing and Information Science in Engineering*, 2016, 16 (4), 041006.
107. C. Prakash, H.K. Kansal, B.S. Pabla and S. Puri. Potential of powder mixed electric discharge machining to enhance the wear and tribological performance of β-Ti implant for orthopedic applications. *Journal of Nanoengineering and Nanomanufacturing*, 2015, 5 (4), 261–269.
108. C. Prakash, S. Singh, K. Verma, S.S. Sidhu and S. Singh. Synthesis and characterization of Mg-Zn-Mn-HA composite by spark plasma sintering process for orthopedic applications. *Vacuum*, 2018, 155, 578–584.
109. C. Prakash, S. Singh, C.I. Pruncu, V. Mishra, G. Królczyk, D.Y. Pimenov and A. Pramanik. Surface modification of Ti-6Al-4V alloy by electrical discharge coating process using partially sintered Ti-Nb electrode. *Materials*, 2019, 12 (7), 1006.
110. C. Prakash, H.K. Kansal, B.S. Pabla and S. Puri. On the influence of nanoporous layer fabricated by PMEDM on β-Ti implant: biological and computational evaluation of bone-implant interface. *Materials Today: Proceedings*, 2017, 4 (2), 2298–2307.
111. C. Prakash, H.K. Kansal, B.S. Pabla and S. Puri. Effect of surface nano-porosities fabricated by powder mixed electric discharge machining on bone-implant interface: an experimental and finite element study. *Nanoscience and Nanotechnology Letters*, 2016, 8 (10), 815–826.
112. C. Prakash, H.K. Kansal, B.S. Pabla and S. Puri. To optimize the surface roughness and microhardness of β-Ti alloy in PMEDM process using non-dominated sorting genetic algorithm-II. *In 2015 2nd International Conference on Recent Advances in Engineering & Computational Sciences (RAECS)*, December 2015 (pp. 1–6). IEEE.
113. C. Prakash, H.K. Kansal, B.S. Pabla and S. Puri. Potential of silicon powder-mixed electro spark alloying for surface modification of β-phase titanium alloy for orthopedic applications. *Materials Today: Proceedings*, 2017, 4 (9), 10080–10083.
114. A. Pramanik, A.K. Basak and C. Prakash. Understanding the wire electrical discharge machining of Ti6Al4V alloy. *Heliyon*, 2019, 5 (4), e01473.
115. A. Pramanik, M.N. Islam, A.K. Basak, Y. Dong, G. Littlefair and C. Prakash. Optimizing dimensional accuracy of titanium alloy features produced by wire electrical discharge machining. *Materials and Manufacturing Processes*, 2019, 34 (10), 1083–1090.
116. C. Prakash, S. Singh, M. Singh, P. Antil, A.A.A. Aliyu, A.M. Abdul-Rani and S.S. Sidhu. Multi-objective optimization of MWCNT mixed electric discharge machining of Al–30SiC p MMC using particle swarm optimization. In : Singh, S., Sidhu, Bains, P.S., Zitoune, R., Yazdani, M. (eds.) *Futuristic Composites*, 2018 (pp. 145–164). Springer, Singapore.

Machining of Engineering Materials

117. C. Prakash, H.K. Kansal, B.S. Pabla and S. Puri. Research on the formation of nano-porous biocompatible layer on Ti-6Al-4V implant by powder mixed electric discharge machining for biomedical applications. In *International Conference*, March 2016, NanoSciTech, Punjab University, Chandigarh.

118. P. Antil, S. Singh, A. Manna and C. Prakash. Electrochemical discharge drilling of polymer matrix composites. In: Singh, S., Sidhu, Bains, P.S., Zitoune, R., Yazdani, M. (eds.) *Futuristic Composites*, 2018 (pp. 223–243). Springer, Singapore.

119. P. Antil, S. Singh, S. Singh, C. Prakash and C.I. Pruncu. Metaheuristic approach in machinability evaluation of silicon carbide particle/glass fiber–reinforced polymer matrix composites during electrochemical discharge machining process. *Measurement and Control*, 2019, 52 (7–8), 1167–1176.

120. Aliyu A.A. et al. (2019) Synthesis and Characterization of Bioceramic Oxide Coating on Zr-Ti-Cu-Ni-Be BMG by Electro Discharge Process. In: Gapiński B., Szostak M., Ivanov V. (eds) Advances in Manufacturing II. MANUFACTURING 2019. Lecture Notes in Mechanical Engineering. Springer, Cham.

121. A. Malik, S. Pradhan, G.S. Mann, C. Prakash and S. Singh. Subtractive versus hybrid manufacturing. *Encyclopedia of Renewable and Sustainable Materials*, 2019, 5, 474–502.

122. C. Prakash, S. Singh, S. Sharma, J. Singh, G. Singh, M. Mehta, M. Mittal and H. Kumar. Fabrication of low elastic modulus Ti50Nb30HA20 alloy by rapid microwave sintering technique for biomedical applications. *Materials Today: Proceedings*, 2020, 21, 1713–1716.

123. C. Prakash, S. Singh, S. Ramakrishna, G. Królczyk and C.H. Le. Microwave sintering of porous Ti–Nb-HA composite with high strength and enhanced bioactivity for implant applications. *Journal of Alloys and Compounds*, 2020, 824, 153774.

124. C. Prakash and S. Singh. On the characterization of functionally graded biomaterial primed through a novel plaster mold casting process. *Materials Science and Engineering: C*, 2020, 110, 110654.

125. C. Prakash, S. Singh, S. Sharma, H. Garg, J. Singh, H. Kumar and G. Singh. Fabrication of aluminium carbon nano tube silicon carbide particles based hybrid nano-composite by spark plasma sintering. *Materials Today: Proceedings*, 2019, 21, 1637–1642.

126. C. Prakash, S. Singh, A. Basak, G. Królczyk, A. Pramanik, L. Lamberti and C.I. Pruncu. Processing of Ti50Nb50– xHAx composites by rapid microwave sintering technique for biomedical applications. *Journal of Materials Research and Technology*, 2020, 9 (1), 242–252.

2 Optimization of Machining Parameters of High-Speed Toolpath to Achieve Minimum Cycle Time for Ti-6Al-4V

Ganesh Kakandikar
Dr. V. D. Karad MIT World Peace University

Rahul Dhage
Unicorn CNC Technologies

Omkar Kulkarni
Dr. V. D. Karad MIT World Peace University

Vilas Nandedkar
SGGS Institute of Engineering and Technology

CONTENTS

2.1 Introduction to High-Speed Machining..24
2.2 Experimental Setup 26
    2.2.1 Selection of Material..26
    2.2.2 Job Setup on Machine...26
    2.2.3 Machine Specifications...28
    2.2.4 High Helix Cutter 28
    2.2.5 Machining Strategy: Dynamic Cutting Strategy from
            Mastercam Software...29
    2.2.6 Selection of Parameters 30
    2.2.7 Design of Experimentation 31
    2.2.8 Results from Response Surface Methodology.........................31
2.3 Results and Discussions..33
    2.3.1 Effect of Process Parameters on the Spindle Load 34
    2.3.2 Effect of the Process Parameters on Cycle Time 37
2.4 Conclusion 41
References..43

2.1 INTRODUCTION TO HIGH-SPEED MACHINING

High-speed machining (HSM) [7], over the past few years, has gained great importance and visibility, as current technological advances have supported its implementation. Due to various improvements in cutting tool and machine technology, HSM [8] has become an economical manufacturing process for production of various parts with increased high precision and surface quality. Until lately, high-speed manufacturing was applied in machining of the aluminum alloys for production of only complicated parts, which are utilized in the aircraft industry. This was the technology that was effectively applied with substantial improvements in machine tools, controllers, and spindles. Recently, with the development of cutting tool technologies, HSM [10] has also been engaged in machining alloy steels for making dies/molds used in the production of a wide range of automotive components, as well as plastic molding parts. HSM is defined as machining at significantly high cutting speeds and feed rates compared with those in old-fashioned methods. HSM [17] is achieved in high material removal rates, lower cutting forces, improved part precision, and reduced lead times. The dissimilarity between conventional machining and HSM is based on the workpiece material being machined, type of cutting operation, and the cutting tool used. High-speed dynamic milling toolpaths utilize the entire flute length of their cutting tools to achieve great efficiency in milling. They are designed to maximize material removal [18,19] while minimizing tool wear. Additional benefits gained by using high-speed dynamic milling toolpaths include (i) tool burial avoidance, (ii) minimum heat buildup, and (iii) better chip evacuation. Following are high-speed dynamic milling toolpaths available in Mastercam [6].

A. **Dynamic area mill:** This strategy machines pockets using one or more chains to drive the toolpath. The chain that encompasses the largest 2D area contains the toolpath; all subsequent chains are considered islands. As shown in Figure 2.1, dynamic area mill normally would machine with a pocket toolpath [9].
B. **Dynamic core mill:** This strategy machines open pocket shapes or standing core shapes using the outmost chain as the stock boundary. The tool moves freely outside of the area. As shown in Figure 2.2, we can use dynamic core mill on parts where we need to face an area, generally on the outside of a part [14].

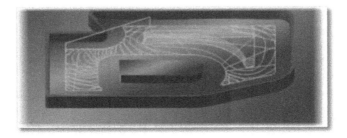

FIGURE 2.1 Dynamic area mill, Mastercam.

High-Speed Machining of Ti-6Al-4V

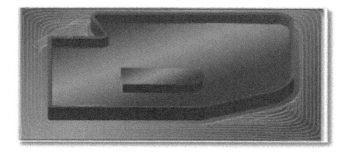

FIGURE 2.2 Dynamic core mill, Mastercam.

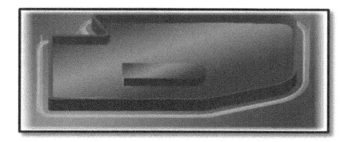

FIGURE 2.3 Dynamic contour.

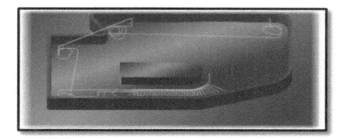

FIGURE 2.4 Dynamic rest mill, Mastercam.

C. **Dynamic contour:** This operation mills material off walls supporting, closed or open chains. The tool motion converts to dynamic, in areas where the tool would bind or encounter too much material. As shown in Figure 2.3, we can use dynamic counter on walls that contain small radii.

D. **Dynamic rest mill:** As shown in Figure 2.4, dynamic rest mill targets material left from previous operations. Dynamic motions are used to

26 Advanced Manufacturing and Processing Technology

rapidly remove material left by previous operations or by a specified tool size. Dynamic rest mill can be used when the roughing tools are used to leave material in corners or narrow part areas.

2.2 EXPERIMENTAL SETUP

2.2.1 Selection of Material

The workpiece material used for this experiment is Ti-6Al-4V. The workpiece is of cylindrical shape, with dimensions with diameter of 25 mm and length of 500 mm, divided in pieces of 150 mm. Chemical composition and physical properties of material are given in Table 2.1. The workpiece is round bar, which is shown in Figure 2.5 [5].

2.2.2 Job Setup on Machine

The whole experiments were carried at Sterling Precession Engineering, Bhivandi. The experimental setup is shown in Figure 2.6; the equipment utilized is vertical machining center (VMC) DMG DMU 650 V machine with Heidenhain controller. The utilized tool is solid carbide tool with the diameter of $\Phi 12$ mm. The workpiece material is Ti-6Al-4V. All operations designed were open pocketing milling

TABLE 2.1
Titanium Ti-6Al-4V Properties

Physical Properties

Density	4.43 g/cc

Mechanical Properties

Hardness, Brinell	379
Hardness, Knoop	414
Hardness, Rockwell C	41
Hardness, Vickers	396
Tensile strength, ultimate	1170 MPa
Tensile strength, yield	1100 MPa
Elongation at break	10%
Modulus of elasticity	114 GPa
Compressive yield strength	1070 MPa
Notched tensile strength	1550 MPa
Ultimate bearing strength	2140 MPa
Bearing yield strength	1790 MPa
Poisson's ratio	0.33
Charpy impact	23 Jt
Fatigue strength	160 MPa
Fatigue strength	700 MPa
Fracture toughness	43 Pa.m$^{1/2}$

High-Speed Machining of Ti-6Al-4V

FIGURE 2.5 Titanium bar. (Courtesy: Sterling Precision Components.)

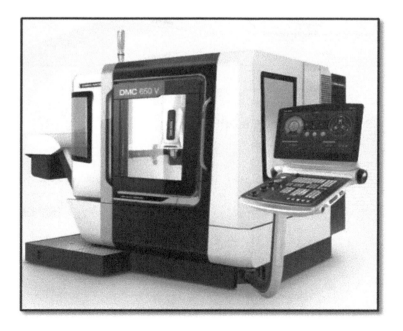

FIGURE 2.6 DMG DMU 650 V machine with Heidenhain controller. (Courtesy: DMG Catalogue.)

operations. Parameters considered for the experiment were spindle speed, feed rate, stepover, and depth of cut as shown in Table 2.2.

2.2.3 MACHINE SPECIFICATIONS

The machine used for cutting trials is DMG DMU 650 V as shown in Figure 2.6 with the following specifications given in Table 2.3. The machine has rigid design with fixed table and elevated X-slide. The machine bed is made of mineralic casting with stand made of cast iron. Tool magazine is of disk type with 20 pockets. The tools can be loaded from front during machining also. SpeedMASTER spindle with 20,000 rpm is available in basic version.

2.2.4 HIGH HELIX CUTTER

Helix angle, as shown in Figure 2.7, is the angle formed by a line tangent to the helix and a plane through the axis of the cutter or the cutting-edge angle that a helical cutting edge makes with a plane containing the axis of a cylindrical cutter [15].

TABLE 2.2
Process Parameters with Levels

Factor	Name	Low Level	Mean Level	High Level
A	Feed	636	912	1188
B	Speed	3183	3713.5	4244
C	Depth of cut	5	10	15
D	Stepover	0.84	1.02	1.2

TABLE 2.3
Machine Specifications

Working Area

X-axis	650 mm
Y-axis	520 mm
Z-axis	475 mm

Feed

Feed force	6.5 KN
Feed rate	42,000 mm/min
Rapid traverse	42 m/min
Acceleration	5 m/s^2

Spindle (Standard)

Maximum speed range	14,000 rpm
Drive power (40%/100% DC)	20.3/14.5 Kw
Torque (40%/100% DC)	121/84 Nm

High-Speed Machining of Ti-6Al-4V

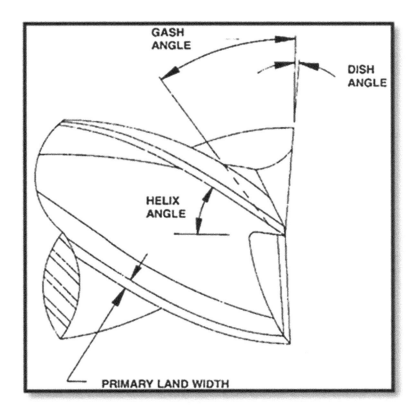

FIGURE 2.7 Helix angle as related to end mills.

Helix can be either right handed or left handed. There are also end mills that have hi-helix and low-helix. Low-helix end mills generally have a 35° or less helix angle. Hi-helix end mills have a 35° or greater helix angle. Some tools used for roughing and finishing may use a 38° helix angle. This gives us the balanced benefit of a low enough helix to reduce chatter and enough helix to obtain good finish passes.

2.2.5 Machining Strategy: Dynamic Cutting Strategy from Mastercam Software

Mastercam Dynamic Motion toolpaths deliver powerful benefits while helping you get most out of any machine. These 2D and 3D high-speed dynamic milling toolpaths utilize the entire flute length of their cutting tools to achieve great efficiency in milling. They are designed to maximize material removal while minimizing tool wear. Additional benefits you gain by using high-speed dynamic milling toolpaths include tool burial avoidance, minimum heat buildup, and better chip evacuation. Figures 2.8 and 2.9 show toolpath generated in Mastercam.

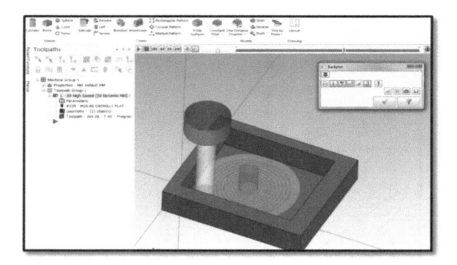

FIGURE 2.8 High-speed toolpath in Mastercam. (Courtesy: Onward Technologies Ltd.)

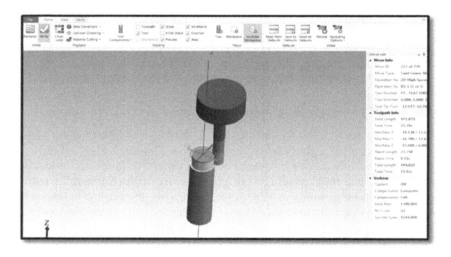

FIGURE 2.9 3D simulation in Mastercam. (Courtesy: Onward Technologies Ltd.)

2.2.6 Selection of Parameters

For HSM, the parameters such as feed, spindle speed, stepover, depth of cut, coolant speed, clamping, cutting tool, tool holder, material hardness, and tool diameter can be treated as machining variables. For design of experimentation, four variables feed, spindle speed, stepover, and depth of cut are identified, and other parameters are treated as constant. So function equation can be written as $Y = f$ (feed, spindle speed, stepover, depth of cut) + error [1].

High-Speed Machining of Ti-6Al-4V

2.2.7 Design of Experimentation

The software utilized for this project is Design Expert 10. Design Expert offers a very deep and comparative test, screening, characterization, optimization, robust parameter design, mixture designs, and combined designs. The statistical importance of these factors is recognized with analysis of variance (ANOVA) [2]. Designs of experimentation (DOEs) are carried out in engineering, agriculture, pharmaceutical, chemical, automotive, electronics, hard goods manufacturing, and so on. Conventionally, experimentation has been done in random, one-factor-at-a-time (OFAT) manner. This method is inefficient and very often yields misleading results. On the other hand, factorial designs are basic type of DOE [3] and require only a nominal number of runs, yet they allow you to identify connections in your process [11]. For experimentation 5, levels of each of the four variables are selected, and central composite design–type response surface methodology (RSM) [16] design is done. Table 2.4 shows the summary of the design planned. From Design Expert software, a run sheet was generated, which shows 25 experiments to be conducted for each spindle load and cycle time responses [12].

Experiments [13] were carried out as per the run table as shown in Table 2.5, and responses after milling machining are noted [4]. The same are entered in the run sheet, and results were ploted from Design Expert software.

2.2.8 Results from Response Surface Methodology

Polynomial equations were obtained from Design Expert software as in the following:

$$
\begin{aligned}
\textbf{SPINDLE LOAD} = {} & -27.4439 + 0.00442822 * \text{FEED} \\
& + 0.00418649 * \text{SPEED} + -1.22797 * \\
& \text{DEPTH OF CUT} + 38.7282 * \text{STEP} \\
& \text{OVER} + -3.14585e - 006 * \text{FEED} * \\
& \text{SPEED} + 0.000311904 * \text{FEED} * \text{DEPTH} \\
& \text{OF CUT} + 0.00592746 * \text{FEED} * \text{STEP} \\
& \text{OVER} + 3.6155e - 005 * \text{SPEED} * \text{DEPTH OF} \\
& \text{CUT} + 0.00763851 * \text{SPEED} * \text{STEP OVER} \\
& + 0.171155 * \text{DEPTH OF CUT} * \text{STEP OVER} \\
& + 2.39261e - 006 * \text{FEED}^2 + -1.06803e - 006 * \\
& \text{SPEED}^2 + 0.0564184 * \text{DEPTH OF CUT}^2 \\
& + -34.291 * \text{STEP OVER}^2
\end{aligned}
\tag{2.1}
$$

TABLE 2.4
Input Parameters with Levels

Factor	Name	Units	Type	Subtype	Mini.	Maxi.	Coded	Values	Mean	Std. Dev.
A	Feed	mm/min	Numeric	Discrete	636	1188	False	1.00=118	917.52	223.5
B	Speed	Rpm	Numeric	Discrete	3183	4244	False	1.00=424	3713.6	433.2
C	Depth of cut	Mm	Numeric	Discrete	5	15	False	1.00=15	10.2	4.077
D	Stepover	Mm	Numeric	Discrete	0.84	1.2	False	1.00=1.2	1.0092	0.143

Response	Name	Units	Obs	Analysis	Mini.	Maxi.	Mean	Std. Dev.	Ratio	Trans	Model
R1	Spindle load	%	25	Polyno.	3	13	6.56	2.887	4.3333	None	Quad.
R2	Cycle time	s	25	Polyno.	31	68	46.04	12.129	2.1936	None	Linear

High-Speed Machining of Ti-6Al-4V

TABLE 2.5
Run Sheet from Design Expert 10 Software

	Factor 1	Factor 2	Factor 3	Factor 4	Response 1	Response 2
			C: Depth of			
Run	A: Feed	B: Speed	Cut	D: Stepover	Spindle Load	Cycle Time
	mm/min	rpm	mm	Mm	%	s
1	1188	3714	5	0.84	5	35
2	912	3714	5	1.02	5	47
3	636	3714	10	0.84	3	66
4	1050	3448	10	1.11	8	34
5	1188	3183	5	1.2	5	31
6	1050	3448	10	1.11	6	39
7	636	3183	15	1.2	5	56
8	912	3183	7.5	0.84	3	48
9	1050	3448	10	1.11	8	35
10	636	3183	15	0.84	6	68
11	1188	4244	15	0.84	10	39
12	1188	3183	15	0.84	10	38
13	912	4244	7.5	0.93	5	45
14	912	3183	7.5	0.84	3	52
15	912	3714	15	1.02	10	43
16	636	4244	15	1.2	10	59
17	1188	4244	10	1.02	8	33
18	1188	4244	10	1.02	8	33
19	912	4244	5	1.2	6	36
20	912	3714	15	1.02	10	43
21	636	3183	5	1.02	3	57
22	636	4244	15	0.93	8	62
23	1188	3979	15	1.2	13	31
24	636	4244	5	0.84	3	67
25	636	3714	7.5	1.2	3	54

$$\text{CYCLE TIME} = 117.381 + -0.0497481 * \text{FEED} + -0.000181085 *$$

$$\text{SPEED} + 0.186781 * \text{DEPTH OF CUT} + -26.6832 * \quad (2.2)$$

$$\text{STEP OVER}$$

2.3   RESULTS AND DISCUSSIONS

RSM designates the area in the design province where the process is likely to give expected results of multiple responses, which involves first building an appropriate response surface model for each and every response and trying to find operating

34 Advanced Manufacturing and Processing Technology

conditions, which in some sense optimize all responses or at least keep them in the preferred range. The various parameters in milling such as the cutting speed, depth of cut, stepover, the feed rate, and so on are the examined parameters that distress the spindle load and the cycle time of the milled parts. Table 2.6 signifies the experimental results with actual process parameters. Spindle load and time required for machining is recorded from controller as shown in Figure 2.10 while doing machining.

2.3.1 EFFECT OF PROCESS PARAMETERS ON THE SPINDLE LOAD

Design Expert 10.0.03 that has been used for all the calculation is shown in Table 2.6. Based on the ANOVA method, Table 2.6 depicts the effects of the process variables and the connections based on the quadratic model for spindle load.

TABLE 2.6

Run Sheet from Design Expert 10 Software for Spindle Load

ANOVA for Response Surface Quadratic Model

Analysis of Variance Table (Partial Sum of Squares—Type III)

Source	Sum of Squares	df	Mean Square	F Value	p-value Prob > F	
Model	195.01	14	13.93	27.03	<0.0001	Significant
A—Feed	41	1	41	79.56	<0.0001	Significant
B—Speed	9.27	1	9.27	17.99	0.0017	Not significant
C—Depth of cut	82.52	1	82.52	160.14	< 0.0001	Significant
D—Stepover	8.17	1	8.17	15.85	0.0026	Not significant
AB	2.03	1	2.03	3.93	0.0755	Not significant
AC	1.85	1	1.85	3.59	0.0875	Not significant
AD	0.91	1	0.91	1.76	0.2139	Not significant
BC	0.092	1	0.092	0.18	0.6822	Not significant
BD	4.61	1	4.61	8.94	0.0136	Not significant
CD	0.24	1	0.24	0.46	0.5148	Not significant
A^2	0.11	1	0.11	0.21	0.6539	Not significant
B^2	0.37	1	0.37	0.72	0.4158	Not significant
C^2	8.21	1	8.21	15.93	0.0026	Not significant
D^2	3.81	1	3.81	7.4	0.0216	Not significant
Residual	5.15	10	0.52			
Lack of fit	2.49	5	0.5	0.93	0.5297	not significant
Pure error	2.67	5	0.53			
Correct total	200.16	24				

Standard deviation	0.72		R-squared	0.9743
Mean	6.56		Adj. R-squared	0.9382
C.V. %	10.94		Pred. R-squared	0.6611
PRESS	67.83		Adeq. precision	18.887

High-Speed Machining of Ti-6Al-4V

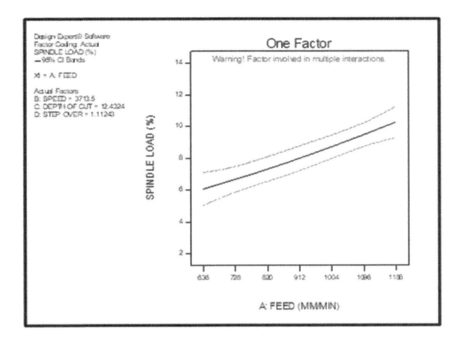

FIGURE 2.10 Main factor plots on spindle load with varied feed rate.

This model was developed for the 95% confidence level. The model F value 27.03 implies that the model is significant. There is only 0.01% chance that model "F value" could exist due to noise. Values of "Prob>F" less than 0.0005 indicate that model terms are significant. Values greater than 0.0005 indicate that the model terms are not significant. The "Pred. R-squared" of 0.6611 is in reasonable agreement with the "Adj. R-squared" of 0.9382. "Adeq. precision" measures the signal to noise ratio. A ratio greater than 4 is desirable. The ratio of 18.887 indicates an adequate signal. This model can be used to navigate the design space. Final Equation 2.3 in terms of actual factors:

SPINDLE LOAD = $-27.4439 + 0.00442822 * \text{FEED} + 0.00418649 *$
$\text{SPEED} + -1.22797 * \text{DEPTH OF CUT} + 38.7282 *$
$\text{STEP OVER} + -3.14585e - 006 * \text{FEED} * \text{SPEED}$
$+ 0.000311904 * \text{FEED} * \text{DEPTH OF CUT}$
$+ 0.00592746 * \text{FEED} * \text{STEP OVER} + 3.6155e - 005$
$* \text{SPEED} * \text{DEPTH OF CUT} + 0.00763851 * \text{SPEED}$
$* \text{STEP OVER} + 0.171155 * \text{DEPTH OF CUT} *$
$\text{STEP OVER} + 2.39261e - 006 * \text{FEED}^2 + -1.06803e$
$-006 * \text{SPEED}^2 + 0.0564184 * \text{DEPTH OF CUT}^2$
$+ -34.291 * \text{STEP OVER}^2$

(2.3)

The main factor plots on spindle load with varied feed rate from 636 to 1188 mm/min are presented in Figure 2.10.

At spindle speed = 3713.5 rpm, depth of cut = 12.4324 mm and stepover = 1.1124 as shown in Figure 2.10. The main factor plots on spindle load with varied spindle speed from 3183 to 4244 rpm at feed rate = 912 mm/min, depth of cut = 12.43 mm, and stepover = 1.11243 mm are shown in Figure 2.11.

The main factor plots on spindle load with varied depth of cut from 5 to 15 mm at feed rate = 0912 mm/min, spindle speed = 3713.5 rpm, and stepover 1.11243 mm are shown in Figure 2.12. The main factor plots on spindle load with varied stepover from 0.84 to 1.2 mm at feed rate = 0912 mm/min, spindle speed = 3713.5 rpm, and depth of cut 12.4324 mm are shown in Figure 2.13. A contour plot plays a very important role in the study of response surface. From the examination of the contour plot and response surface, it is observed that spindle load increases from 4% to 12%, with depth of cut from 5 to 15 mm with feed rate from 636 to 1188 mm/min.

Figures 2.14 and 2.15 indicate the residual Vs predicted values. From the figure, it is clear that all the data points are following the straight line. Thus, it can be stated that the data are normally distributed. All the actual values are following the predicted value and thus declaring that model assumptions are correct and within the limits.

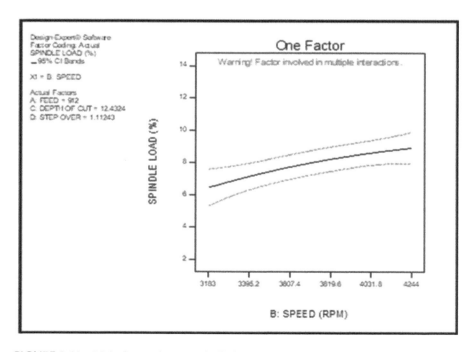

FIGURE 2.11 Main factor plots on spindle load with varied spindle speed.

High-Speed Machining of Ti-6Al-4V

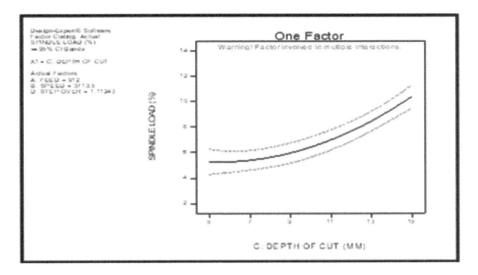

FIGURE 2.12 Main factor plots on spindle load with varied depth of cut.

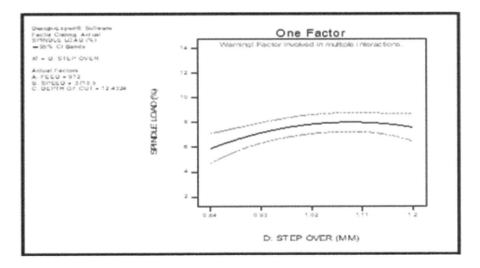

FIGURE 2.13 Main factor plots on spindle load with varied stepover.

2.3.2 Effect of the Process Parameters on Cycle Time

The model F value of 112.85 indicates that the model is important. There is only a chance of 0.01% that "model F-value" this large could occur due to noise. By

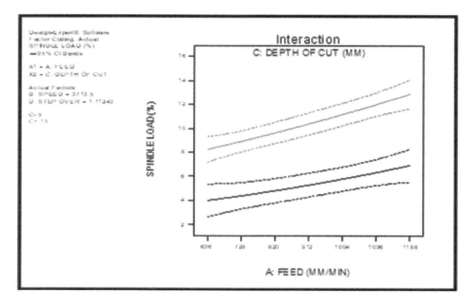

FIGURE 2.14 Interaction plots with feed rate and depth of cut.

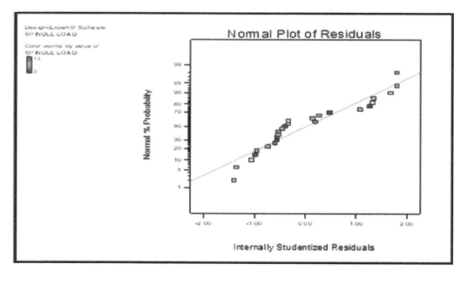

FIGURE 2.15 Normal % probabilities.

High-Speed Machining of Ti-6Al-4V

the checking F values and P values, it is flawless that values of "prob.>F" less than 0.005 indicate that model terms are important. In this model, the term A (feed rate) and term B (speed) have the most significant effect on cycle time as shown in Table 2.7. Values greater than 0.0001 indicate that the model terms are not significant. The "Pred. R-squared" of 0.9309 is in sensible contract with the "Adj. R-squared" of 0.9490. "Adj. R-squared" processes the signal to noise ratio. A ratio greater than 4 is desirable. The ratio 31.81 indicates an adequate signal. This model can be used to navigate the design space. The final Equation 2.4 in terms of actual factors:

$$\mathbf{CYCLE\ TIME} = 117.381 + -0.0497481 * \mathrm{FEED} + -0.000181085$$
$$* \mathrm{SPEED} + 0.186781 * \mathrm{DEPTH\ OF\ CUT}$$
$$+ -26.6832 * \mathrm{STEP\ OVER} \tag{2.4}$$

The main plot on cycle time with varied feed rate from 636 to 1188 mm/min at spindle speed = 3713.50 rpm, depth of cut = 10 mm, and stepover = 1.02 is shown in Figure 2.16, and cycle time with varied stepover from 0.84 to 1.2 mm with spindle speed from 3713.5 rpm at feed rate 912 mm/min and depth of cut 15 mm is shown in Figure 2.17.

TABLE 2.7
Run Sheet from Design Expert 10 Software for Cycle Times

ANOVA for Response Surface Linear Model

Analysis of Variance Table (Partial Sum of Squares—Type III)

Source	Sum of Squares	df	Mean Square	F Value	p-value Prob > F	
Model	3381.16	4	845.29	112.85	<0.0001	Significant
A—Feed	2956.18	1	2956.18	394.67	<0.0001	Significant
B—Speed	0.15	1	0.15	0.02	0.8902	Not significant
C—Depth of cut	13.84	1	13.84	1.85	0.1891	Not significant
D—Stepover	348.71	1	348.71	46.56	<0.0001	Significant
Residual	149.8	20	7.49			
Lack of fit	127.8	15	8.52	1.94	0.24	Not significant
Pure error	22	5	4.4			
Correct total	3530.96	24				

Standard deviation	2.736830926	R-squared	0.957573898	
Mean	46.04	Adj. R-squared	0.949088677	
C.V. %	5.944463349	Pred. R-squared	0.930983353	
PRESS	243.6950188	Adeq. precision	31.810739	
−2 Log likelihood	115.7083706	BIC	131.8027497	
		AICc	128.8662653	

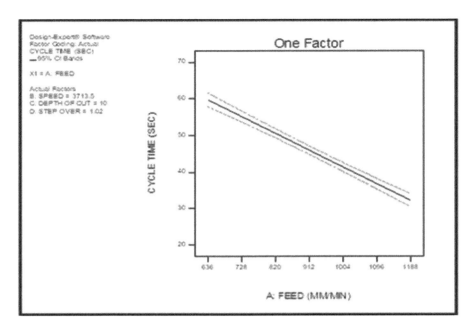

FIGURE 2.16 Cycle time with varied feed rate.

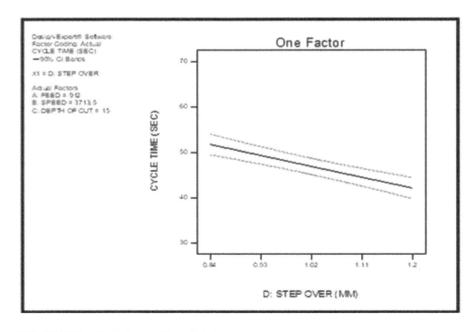

FIGURE 2.17 Cycle time with varied stepover.

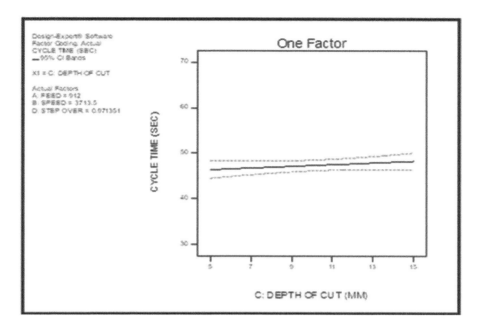

FIGURE 2.18 Cycle time with varied depth of cut.

Varied depth of cut from 0.10 to 0.50 mm at feed rate = 275 mm/min and spindle speed = 1350 rpm is shown in Figure 2.18. Cycle time with varied speed from 3183 to 4244 rpm at feed rate = 912 mm/min at depth of cut 15 mm and stepover 0.97 mm is shown in Figure 2.19. By generating response surface analysis, it is easy to characterize the shape of the surface and locate the optimum with reasonable precision from the examination of the contour plots and response surface, and it is observed that cycle time decreases from 60 to 40 s with feed rate from 636–1188 mm/min and 3183–4244 rpm.

2.4 CONCLUSION

The spindle load and cycle time by mathematical model are 4.61% and 31 s, respectively, whereas the spindle load and cycle time of the experimental result are 5% and 33 s, respectively. Tables 2.8 and 2.9 show the error percentage for experimental validation of the developed models for responses with optimal parameter setting during the milling of the Ti-6Al-4V on vertical machining center (VMC). From the analysis of Table 2.9, it can be observed that the calculated error is small. The error between experimental and predicted values for spindle load and cycle time lies within the 1% and 6.06%, respectively. Hence the aforementioned experimental result confirms excellent reproducibility of the experiment conclusions.

In this study, the spindle load and cycle time in parameter optimization in VMC for Ti-6Al-4V were modeled and analyzed through RSM. Feed rate, spindle speed,

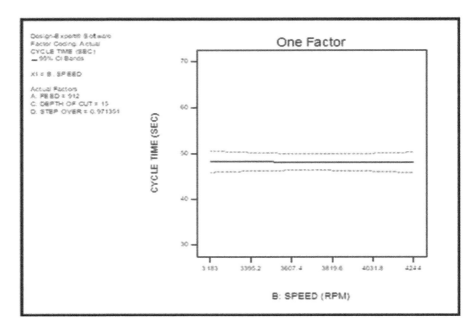

FIGURE 2.19 Cycle time with varied speed.

TABLE 2.8
Multioptimal Parameter Settings for Spindle Load and Cycle Time

Parameters	Units	Optimal Parameters
Feed rate	mm/min	1100
Spindle speed	rpm	3183
Depth of cut	mm	5
Stepover	mm	1.19

TABLE 2.9
Experimental Validations of the Developed Models with Optimal Parameters

Responses	Predicted Value	Experimental Value	Error (%)	Desirability
Spindle load (%)	4.61	5	7.8	0.85
Cycle time	31	33	6.06	0.99

High-Speed Machining of Ti-6Al-4V

depth of cut, and stepover have been employed to carry out the experimental study. Summarizing the main features, the following conclusions could be drawn:

1. The predicted values lie very close to the experimental values, with 5% for spindle load and 33 s for cycle time.
2. The error between the experimental values and predicted values at the optimal combination of parameters for spindle load and cycle time lies within 7.8% and 6.06%, respectively.
3. From multiresponse optimization, the optimal combination of parameters setting is feed rate = 1100 mm/min, spindle speed = 3183 rpm, depth of cut = 5 mm, and stepover = 1.19 mm for achieving the required minimum spindle load and minimum cycle time.

REFERENCES

1. A. Mahamani, Influence of process parameters on cutting force and surface roughness during turning of AA2219-TiB2/ZrB2 in-situ metal matrix composites, *Procedia Materials Science*, Vol. 6, 2014, pp 1178–1186.
2. A. Qasim, S. Nisar, A. Shah, M. S. Khalid, M. A. Sheikh, Optimization of process parameters for machining of AISI-1045 steel using Taguchi design and ANOVA, *Simulation Modelling Practice and Theory*, Vol. 59, 2015, pp 36–51.
3. A. K. Sahoo, S. Pradhan, Modeling and optimization of Al/SiCp MMC machining using taguchi approach, *Measurement*, Vol. 46, 2014, pp 3064–3072.
4. A. M. Zain, H. Haron, S. N. Qasem, S. Sharif, Regression and ANN models for estimating minimum value of machining performance, *Applied Mathematical Modelling*, Vol. 36, 2012, pp 1477–1492.
5. F. Akbar, P. T. Mativenga, M. A. Sheikh, An evaluation of heat partition in high-speed turning of AISI/SAE 4140 steel with uncoated and tin-coated tools, *Proceedings of the Institution of Mechanical Engineers, Part B: Journal of Engineering Manufacture*, Vol. 222, 2008, pp 759–771.
6. H. Oktema, T. Erzurumlub, H. Kurtaranb, Application of response surface methodology in the optimization of cutting conditions for surface roughness, *Design and Analysis of Experiments*, Vol. 32, 2003, pp 210–219, John Wiley and Sons.
7. H. Schulz, T. Moriwaki, High-speed machining, *Annals of the CIRP*, Vol. 41, Issue 2, 1992, pp 637–643.
8. A. Hamdan, A. A. D. Sarhan, M. Hamdi, An optimization method of the machining parameters in high-speed machining of stainless steel using coated carbide tool for best surface finish, *International Journal of Advanced Manufacturing Technology*, Vol. 58, 2012, pp 81–91.
9. J. Paulo Davim, V. N. Gaitonde, S. R. Karnik, Investigations into the effect of cutting conditions on surface roughness in turning of free machining steel by ANN models, *Journal of Materials Processing Technology*, Vol. 205, 2008, pp 16–23.
10. J. Tlusty, High-speed machining, *Annals of the CIRP*, Vol. 42, Issue 2, 1993, pp 733–738.
11. J. S. Pang, M. N. M. Ansari, O. S. Zaroog, M. H. Ali, S. M. Sapuan, Taguchi design optimization of machining parameters on the CNC end milling process of halloysite nanotube with aluminium reinforced epoxy matrix (HNT/Al/Ep) hybrid composite, *HBRC Journal*, Vol. 10, 2014, pp 138–144.

12. K. Venkatesana, R. Ramanujan, P. Kuppan, Analysis of cutting forces and temperature in laser assisted machining of Inconel 718 using Taguchi method, *Procedia Engineering*, Vol. 97, 2014, pp 1637–1646.
13. L. M. Maiyar, R. Ramanujan, K. Venkatesan, J. Jerald, Optimization of machining parameters for end milling of Inconel 718 super alloy using Taguchi based grey relational analysis, *Procedia Engineering*, Vol. 64, 2013, pp 1276–1282.
14. Mastercam Catalog, *Highspeed Machining*, 2012. www.mastercam.com.
15. N. Masmiati, A. A. D. Sarhan, Optimizing cutting parameters in inclined end milling for minimum surface residual stress–Taguchi approach, *Measurement*, Vol. 60, 2015, pp 267–275.
16. N. Muthukrishnan, J. Paulo Davim, Optimization of machining parameters of Al/Sic-MMC with ANOVA and ANN analysis, *Journal of Materials Processing Technology*, Vol. 209, 2009, pp 225–232.
17. R. C. Dawes, D. K. Aspinwall, A review of ultra-high-speed milling of hardened steels, *Journal of Materials Processing Technology*, Vol. 69, 1997, pp 1–17.
18. S. Tiryaki, A. Malkocoglu, S. Ozsahin, Using artificial neural networks for modeling surface roughness of wood in machining process, *Construction and Building Materials*, Vol. 66, 2014, pp 329–335.
19. T. Kıvak, Optimization of surface roughness and flank wear using the Taguchi method in milling of hadfield steel with PVD and CVD coated inserts, *Measurement*, Vol. 50, 2014, pp 19–28.

3 A Review of Machinability Aspects of Difficult-to-Cut Materials Using Microtexture Patterns

Rahul Sharma and Swastik Pradhan
Lovely Professional University

Ravi Nathuram Bathe
International Advanced Research Centre for
Powder Metallurgy and New Materials (ARCI)

CONTENTS

3.1 Introduction 46
    3.1.1 Judging Machinability 47
        3.1.1.1 Tool Life..47
        3.1.1.2 Power Consumption 47
        3.1.1.3 Surface Finish 47
        3.1.1.4 Chip Form 48
    3.1.2 Difficult-to-Cut Material 49
    3.1.3 Compilation of Machining Technologies 49
    3.1.4 Various Steps Taken to Solve Issues...50
        3.1.4.1 Hot Machining 50
        3.1.4.2 Minimum Quantity of Lubricant 51
        3.1.4.3 Coated Tools 51
        3.1.4.4 High-Speed Machining..52
        3.1.4.5 Flood Cooling 53
        3.1.4.6 Microgrooves 53
3.2 Literature Review 54
3.3 Discussion and Future Work...61
3.4 Conclusions...62
References...62

3.1 INTRODUCTION

Industrialization is a very significant part of any engineering understanding. Machining, which includes a number of processes, converts a raw material into the looked-for shape and product with desired properties. It involves a number of processes, different shapes, and sizes of the products. Gross domestic product (GDP) of many countries is affected by the manufacturing segment. This segment also provides maximum jobs to the people of same and other countries. That is why, "Make in India" program was launched in 2014 in India. Hence, it is required to carry out in-depth research in all manufacturing processes for making it energy efficient along with improvements in surface finish quality. Also a green manufacturing is to be developed for protecting our environment, human lives, and status. It is possible only if tribological studies of tool insert and workpiece interface are done.

In machining, all processes are classified into three principle categories which are milling, turning, and drilling. All other operations come into the subcategories of these.

- In milling operation, the cutting tool rotates so that the cutting edges of tool stand against the workpiece for material removal process. Milling machines are the foremost machine used for these operations.
- Turning operation is the primary method of removing material by moving workpiece of tool against each other. Lathe machines are being widely used for this operation.
- Drilling operations use a rotating cutter to produce holes into the workpieces. Drilling operations are done in drilling machines and milling machines and for some products even on lathes.
- These primary operations have a number of miscellaneous processes such as shaping, boring, sawing, broaching, and so on.

The perfect machining has to meet a number of parameters provided in technical drawings or blueprints of product. After meeting the dimensional accuracy, the desired surface finish is also a challenge in the field manufacturing. The main reason of finishing errors occurs due to use of damaged tool, wrong selection of cutting tool, clamping error, and so on, which results into the irregular surface and cracks on the machined surfaces.

The machinability means the ease with which any product can be machined or undergo various machining processes for removing the material for the desired and satisfactory finish. These methods must be economically enough. The machinability includes less power to cut, good surface finish, less tool wear, and high tool life. Hence, all the manufacturing engineers are keen observers for all parameters to improve the machinability of a process or apparatus.

Machinability can be a challenge to achieve because it has a long list of variables. We can divide these variables into two set of categories: material's condition and physical properties. The condition of material means the factors such as microstructure, hardness, grain size, chemical composition, heat treatment, yield, and tensile strength. Physical properties are elasticity, work hardening, thermal conductivity,

Machinability Aspects of Difficult-to-Cut Materials

and expansion. Besides from these, some other important factors that are important to be considered are material and geometry of cutting tool, operating conditions, and process parameters.

3.1.1 JUDGING MACHINABILITY

When the machinability is to be judged, then the main factors that are to be evaluated are given in the following:

3.1.1.1 Tool Life

It is the time period for which the performance of cutting tool remains well enough giving desired results efficiently. Tool life determination becomes very difficult due to variables such as tool geometry, workpiece material, conditions of machining operations, cutting speed, temperature, and depth of cut. During cutting, less resource intake is required. Yet, abrasive properties, which are the main reason of rapid tool wear, reduce the machinability of material and increase the cost of production. Tool life cannot be taken as machinability index because of its sensitivity to many machine variables [1]. If tools of two different materials are used, then their machinability ranking cannot be compared only on the basis of tool life like a high-speed steel tool, and a sintered carbide tool is being used in two different cases [2].

3.1.1.2 Power Consumption

There are two reasons to include power consumption as a criterion of machinability. First, less power is required if a material is easy to move against cutting tool. Second, machining cost per product in terms of power used is minimum.

It is good to take specific energy of metal as an identification of machinability because it is mainly a property of workpiece material and is insensitive to tool material. By contrast, tool life is strongly dependent on tool material.

3.1.1.3 Surface Finish

Value of surface finish is also an important part of judging the machinability of a metal. Soft and ductile materials always tend to form an edge. But all those materials which machine with a relative high shear zone always tend to minimize the builtup edge effects. Such materials are brass, aluminum alloys, titanium and its alloys, cold-worked steels, and free-machining steels. If surface finish alone were the chosen index of machinability, these latter metals would rate higher than those in the first group. There are many cases in which the surface finish has no importance to measure the machinability of workpiece. Rough cuts manufacturing is the example of these. In other manufacturing, surface finish is not the desired part of output parameters; those will also not have machinability judgment on the basis of surface finish [3].

For various materials, it is very easy to determine the machinability. Surface roughness readings are measured with instrument after machining them under very controlled cutting conditions. These values are having a reverse meaning such as high reading that means not a good surface finish quality and low machinability in return. Sometimes, it is better to take relative readings in materials.

3.1.1.4 Chip Form

Machinability can also be judged on the basis of chip form obtained in machining operation. If a long string type chips are formed, then its machinability rating is low. And if very fine chips in form of powder are obtained, then it means its machinability is better. Those materials would have top rating, which inherently forms nicely broken chips. The process of disposal and handling of chips is little expensive. String-type chips are very difficult to remove off from machine and workpiece and are to the operator. But these are required to be removed from freshly machined surface. Chips formation totally depends on the material of workpiece and machine parameters so the machinability can be improved by taking appropriate method of removing them.

If the rating is given on the basis of ease of chip removal, then it is qualitative in nature. It finds some application in drilling operation, where good chip formation action is required. However, the flogging action of long coils once they are clear of the hole is unwanted.

- **Cutting temperature:** The cutting temperature is the most effective parameter of determining the machinability as it mostly affects tool life and job surface finish also. The main part of heat generated is disposed away through the chips. But as chips are to be removed earlier during machining to obtain good quality surface, it cannot add more effect in reduction of cutting temperature. So the chips must be taken away when maximum part of heat is disposed away through chips so that effect on tool and workpiece can be minimized. The main effects of high cutting temperature on tool are as follows:
  - Reduction in tool life
  - If the tool material is not so strong or hard to withstand the heat generated by them, plastic deformation of cutting edges will occur
  - Fracturing of cutting edges due to thermal shocks.

The main effects of cutting temperature on the workpiece are as follows:

- Distortion in shape and inaccuracy in dimensions due to the expansion and contraction because of high temperature
- Oxidation and burning of surface which may result in the rapid corrosion of workpiece
- Generation of microcracks
- Initiation of tensile residual stresses

No doubt the rise in cutting temperature will reduce the cutting force required for machining by softening the work material and reducing the shear strain, which leads to the reduction in power consumption. But in machining of metals, the tool–chip interface temperature is a prime function of the cutting parameters, and this magnitude of temperature directly affects production of that particular metal. So modern research of reducing the cutting temperature can lead to the improvement of machinability [4].

Machinability Aspects of Difficult-to-Cut Materials

3.1.2 DIFFICULT-TO-CUT MATERIAL

Any material in general which is difficult to machine with existing methods of machining are known as difficult-to-cut material. Stainless steel was also considered as the material of this category when sufficient methods were not developed for its machining. But due to the developments of manufacturing techniques, it is excluded from that list.

The main difficult-to-cut materials are Inconel, titanium, Super Invar, nickel, and Low thermal expansion coefficient material. People from the United States had difficulty finding proper cutting methods of such material.

Super alloys are having superior properties due to which these are one of the high performance materials used in various industries. These properties are high tensile and strength to weight ratio, high compressive strength, high fatigue resistance in sea, lower density, and high resistance to corrosion. Superalloys are also included in this category because of their poor thermal conductivity during machining, low modulus of elasticity of material, and strong tendency of chemical reaction to tool materials at high temperature [5].

3.1.3 COMPILATION OF MACHINING TECHNOLOGIES

During the machining of such materials, the traditional methods cannot give required results. Vibrant factors need to be taken into consideration for desired results. These factors are rigidity and stiffness of cutting tools, repairs and appropriate selection of machining parts, and so on. Superalloys have properties of higher melting temperature due to which these are used in the manufacturing of aerospace and naval machine components. These materials are mainly of four main types: nickel, cobalt, titanium, and iron-based alloys. Titanium, tool steels, stainless steels, hardened steels, and other superalloys are important modern materials to meet the demands of extreme applications. Despite their excellent properties against corrosion and fatigue, high temperature strength, and lower heat sensitivity, they are extremely difficult to machine, resulting in rapid tool wear. By definition, these materials will be machined at high speeds due to the rapid wear risk, even though the actual machining speed could be substantially lower than the typical high processing speeds of aluminum.

In machining, cutting tool plays an important role due to the reason that cutting speed completely depends on the material of cutting tool. That is why, all researchers are continuously trying to find out the tool–workpiece interaction parameters for obtaining better machined surface. To improve machinability of such materials, there is requirement of proper selection of tool, machine tool with all cutting conditions. Figure 3.1 shows the various factors in the field of machining area.

All superalloys or difficult-to-cut materials have high strength at high temperatures. Due to this property, during machining, mostly they produce segmented chips and produce high dynamic forces. These materials have a property of poor heat conductivity. Due to large hardness, it generates high temperatures at the time of machining. All these properties of superalloys create the notch wear, possessing a large depth of cut. Moreover, due to poor thermal property, these metals have high

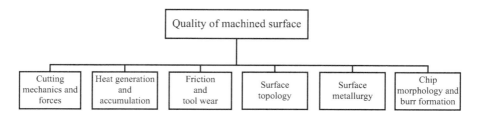

FIGURE 3.1 Main observed issues in machining of various super alloys.

and concentrated temperature at tool/chip interface, which results in increased tool wear and machining cost.

3.1.4 Various Steps Taken to Solve Issues

3.1.4.1 Hot Machining

Different machining approaches have been used to overcome these problems, one of them is called "hot machining." Hot machining is a method to externally heat the workpiece material before or during machining operation. This machining method is very useful to cut difficult-to-machine materials because they are harder and have higher tensile strength than other engineering materials. They become softer when external heating on the workpiece material is applied. As a result, the application of external heat would promote the machinability of material. The application of external heat on workpiece material is simply named as "hot machining." This description is sometimes called as "thermally enhanced machining," "heat-assisted machining," "elevated temperature machining," "plasma-assisted machining," or "laser-assisted machining." In this study, hot machining is preferred due to covering all other descriptions. The advantages of hot machining can be mentioned as follows:

a. Reducing mechanical properties of workpiece materials for easy machining operation
b. Increasing the machinability property of material
c. Increasing to select higher machining parameters (higher cutting speed, depth of cut and feed)
d. Reducing tool wear
e. Increasing productivity
f. Eliminating cutting fluid application
g. Reducing total production cost

These advantages are important when expensive materials are machined and production costs are high. The disadvantages of hot machining, in turn, can be mentioned in the following:

a. Increasing machine tool cost due to external heating technique cost
b. Difficult to control dimensional accuracy due to heat on workpiece material

Machinability Aspects of Difficult-to-Cut Materials

c. Difficult to determine surface finish

d. Require more safety precautions because of heating equipment

3.1.4.2 Minimum Quantity of Lubricant

Many years ago, the concept of MQL (minimum quantity of lubricant) came into play because of some environmental issues and occupational hazards. No doubt the minimization of cutting fluid will give economic benefits by saving cost of lubricant. MQL also save time of machine cleaning, workpiece, and tool mounting. It is also known as "near dry lubrication" or in some cases "microlubrication." In the MQL technique, small quantity of lubricant of a flow rate of 50–500 ml/h toward the cutting zone is used. One or more nozzles are used to provide lubricant in an external system. If we compare the normal lubricating system with MQL, then it has been observed that it is only of one-third to one-fourth in terms of amount of lubricant used. Due to this method, the cost, time, and disposal of cutting heat is affected, which in return effects the machinability [6].

Advantages of MQL system:

Financial advantages

a. Due to controlled supply of coolant, saving of coolant and other resources is there.

b. A higher tool life can be achieved.

c. It reduces the time of machine cycle up to 30%.

d. Purchase and handling cost of lubricant will be reduced.

e. Disposal costs of the coolant will also be reduced.

f. Dry metal chips can be easily recyclable as compared with wet metal chips.

g. Leakage of coolant leads to the accidents, which can be avoided.

h. Operators can be protected from skin diseases.

However, despite the opportunities and benefits that MQL machining can offer, there are still challenges to overcome and some key considerations in implementation:

a. For machining, special tools are required.

b. Investment costs are high.

c. Suitability of the machine is required [7,8].

d. MQL produces a very fine wastage, which is very difficult to be filtered out from lubricants.

e. MQL setup requires a number of changes in machine tool setup and machine area.

3.1.4.3 Coated Tools

Coated tools are already being used on many manufacturing operations. The main concern here is that it must be of such a layer thickness that it possesses very low thermal conductivity, due to which it prevents the ingress of heat into material of tool. The most commonly used materials are TiAlN, TiN, TiAlCrN, and so on. In fact, coated cemented carbides are the widely used cutting tools for major of the manufacturing areas. Moreover, these give high levels of productivity, which make affordable products for the customers. Coated carbide inserts are widely used while working with all ferrous materials. Uncoated carbide inserts are good for machining of nonferrous materials, such as aluminum. These coated tools are having two categories such as chemical vapor deposition (CVD) and physical vapor deposition (PVD). Both have a number of benefits. In thick category, the CVD coatings have a layer of 9–20 microns, and these have high wear resistance. PVD coatings are of thin-type coatings having a layer of 2–3 microns.

Advantages:
a. Coated tools reduce the job setup times and hence reduce manufacturing cost.
b. Different grades of same insert can be used for the machining of different material, which in return reduces the tool inventory.

Disadvantages:
a. The rigidity and power of CNC milling machine must be sufficient for the use of coated tools. These machines should also be able to produce controlled feed rates.

3.1.4.4 High-Speed Machining

In high-speed machining (HSM) process, the engaging time is reduced, and the cuts are narrow. A high feed rate is selected. So a very less cutting force is required, which gives a small tool deflection. This is one of the highly productive and safe processes, which has a constant stock. HSM cannot be of one of the particular processes because it is a combination of a number of processes to be done in less time. It can be defined as follows:

- Operations to be performed at a high cutting speed (vc)
- Operations to be performed with a high spindle speed (n)
- Operations to be performed with a high feed rate (vf)
- Operations to be performed with a high removal rate (Q)

Advantages of HSM:
Radial forces are low in HSM. Spindle forces are also very low, which saves the guide ways, spindle bearings, and ball screws. HSM and axial milling is also a good combination, as the influence on the spindle bearings is less and it allows longer tools with less risk for vibrations. HSM is mainly used in small sized parts in which material removal rate is very low. HSM is used to get fine surface finish as the range of 0.2 microns. It also makes possible to machine a very thin surface. Downmilling tool paths should be used, and the contact time between edge and workpiece must be extremely short to avoid vibrations and deflection of the wall.

Disadvantages of HSM:
In HSM, the maintenance cost increases due to faster wear of ball screws and spindle bearings. HSM cannot be applied without having a complete knowledge of process.

It needs a faster data transfer rate and programming setup. Hence, an experienced and trained operator is required. Proper work plan is required considering all of the precautions and safety conditions. All machine parts and fasteners are required to be checked regularly to avoid any accident. Proper mentioned tools can be used.

3.1.4.5 Flood Cooling

Flood cooling method is commonly used in industries for the purpose of reducing the cutting temperature. It includes the use of soluble oil. It has a huge quantity of lubricant to be supplied to reduce the temperature and to remove chips by a flushing on cutting interface. When a flood of coolant is supplied, it reduces the thermal shock on milling cutter. Ignition of the chips reduces. But this method is not appropriate in machining of titanium. As flood cooling does not have precise and unidirectional application of coolant, machining of titanium has a short contact area so the removed chips act as an obstacle of applying coolant to the cutting area so temperature is not minimized and tool life gets reduced during machining. Moreover, a large quantity of coolant is to be processed for reuse.

3.1.4.6 Microgrooves

There are a number of fields in which the perfect mechanical properties are required. Food processing industries, aeronautical fields, and medical operations have requirement of high surface finish with corrosion-resistant surface. During machining of steels, it is not easy to get required surface finish with traditional methods because of high tensile strength. It has low thermal conductivity [9].

For the enhancement in machinability, many methods such as surface coating or tool geometry modifications have been tried. But that does not produce much effect. Biermann et al. [10] have done experiments using various categories of hard coatings such as TiN, AlCrN, and TiAlN. It was declared that TiAlN and AlTiN coatings give better machining result because of their high hardness (Figure 3.2). Sugihara and Enomoto [11] generated a microtexture to improve the antiadhesion

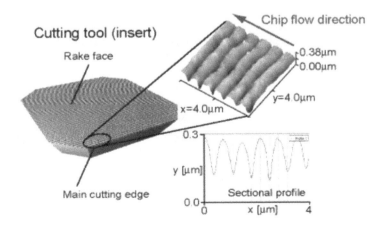

FIGURE 3.2 Cutting tool having microtextured surface [10].

property. They found a major decrease in chip adhesion on face of cutting tool. They have machined the aluminum alloy with microgrooved tools. Their study proved that the modified tool surface with microgrooves is a modern method of enhancing the adhesion resistance. Zhang et al. [12] produced a macro/microsized texture on the surface of coated TiAlN tools. Hence, it results in the rise in antiadhesion-resistant properties. Experiment was performed on AISI 316. It increases the lubrication on the interface of the tool and chips zone. Because of these microtextures, the contact area decreases. Some dimple and channel-type patterns were grooved on the surface of tool, which were then used to machine a plain carbon steel.

They conclude that dry machining with microgroove tool improved the heat transfer capabilities. And the machining in MQL with microgroove tool reduces the cutting temperature, cutting forces, and contact area of tool and chips. If we list down the benefits of textured tools with traditional tools, then we can say the microgrooved tools give an effective reduction in tool–chip interface, provide hydrodynamic lift, and produce effective lubrication, tool wear, cutting forces, and so on

3.2 LITERATURE REVIEW

Various properties, design, simulation, and manufacturing techniques have been studied by a number of researchers by using different problem methods. Some of these have been discussed in the following.

Gente and Hoffmeister [13] explained the chip formation while machining of grade 5 titanium alloy using the cutting speed which varies from 300 to 6000 m/min. They measure the specific cutting forces. They introduced a novel and rapid method in which deceleration was there in a very short distance, which delivers a new data regarding the construction of the segmented chips. During experiments when cutting speed exceeds from 2000 m/min, a new change in the structure of the segmented chips was observed, whereas no change in the specific cutting energy was observed, though it affects the structure of the segmented chips. Ribeiro et al. [14] depicts the effect on titanium material during machining. He declared the best cutting conditions while machining titanium alloys. He concluded that dry turning operation with 90 m/min of cutting speed was the best comparable machining condition. It gives better surface finish. However, when the experiment was done by increasing the cutting speed to 110 m/min, the recorded data of surface finish were increased, but along with this cutting tool, edge deteriorates very quickly. Hua et al. [15] studied the chip morphology while machining of titanium alloy. He concluded that when cutting speed is low, generally discontinuous chips were produced, whereas at high cutting speed, formation of serrated chips takes place. In his study of finite element simulation, the segmented chip generation occurred based on the implicit, Lagrangian nonisothermal rigid viscoplastic model of flow stress. The formation of crack in the chip occurred due to the ductile fracture criteria based on the strain energy. From the simulation result, it was revealed that stress near the tool tip change and the crack propagates toward the free surface of the deformed chip in the shear zone as the cutting speed increased.

Haron and Jawaid [16] performed experiments of rough machining on titanium alloy and studied the surface reliability using uncoated carbide inserts. The cutting speed was taken from 45 to 100 m/min. The applied feed rate was 0.35 to 0.25 mm/

Machinability Aspects of Difficult-to-Cut Materials 55

rev with constant cutting depth of 2 mm. As titanium has less machinability index, the surface finish was not of good quality. The value of surface roughness was in the limit of 6 microns. When SEM was taken, severe microstructure alteration was observed. Umbrello [17] studied the finite element analysis of Ti-6Al-4V during HSM. The study deals with the cutting force, tool wear, segmented chip formed, and the morphology of the chip. Material characterization for FEM analysis was done by using Johnson–cook material model equation. To study the behavior of titanium alloy during HSM, three different sets of material constant were being implemented. The results revealed that the prediction of both principal cutting force and morphology of the segmented chip agreed well with the experimental results. Calamaz et al. [18] have done the numerical simulation of serrated chip in the process of machining of titanium alloy. He performed all simulations and concluded numerical solutions using FORGE 2005 software. The main aim of the new material models is to produce segmented chips during machining of titanium alloy with a wide range of cutting speed and feed. During modeling, it was assumed that the segmentation of chips was induced only by adiabatic shear banding. And it was also fixed that there is no material failure in the primary shear zone. The results were calculated on the basis of strain rate, temperature, strain, and strain soften-ing effect. After that, the results were validated with experiments under same condi-tions. Sun et al. [19] performed experiments at low cutting speed but high feed rate. They found the segmented continuous chips. The slipping angle of continuous chips was 38°. But the same of segmented chips was 55°. During experimentation, it was observed that the peak cyclic force over continuous chips was 1.18 times lesser than the segmented chips. The length of segmented chips does not depend on the cutting speed and depth of cut. Cutting speed increases, and decrease in cutting forces occurs due to the thermal softening of materials. Obikawa et al. [20] proposed the manufac-turing of microtextured tool using laser of femosecond machine. Milling experiments were carried on aluminum alloy. Results proved that a microgrooved surface endorsed the antiadhesive effects, but still there was a problem of adhesion in all experiments. They studied the different ways of improving antiadhesive effects. They found that the microtextured tool expressively improved the lubricity and antiadhesive property. Silva et al. [21] performed experiments taking dry as well as wet conditions. The measur-ing parameters were wear in tool and changes in surface quality. Supply of fluids was controlled by reduced flow rate, MQL, and flooding method. They declared that longer values and higher material removal rate were obtained when experiments are per-formed with reduced flow rate. Doing experiments under these conditions prevents the chipping. Schulze and Zanger [22] had determined the material model parameter of Johnson cook model for titanium alloy by performing split Hopkinson bar tests. They have done the simulation of cutting tool and workpiece taking constant material to pre-dict the surface veracity of the finished surface. It was concluded that surface integrity depends upon the material model and the value of the material constant. Strain rate and cutting temperature was showing opposing mechanisms with the variation of the cutting velocity, and depth of the plastic deformation was rapidly increased. Sahoo et al. [23] have done the conventional casting process and developed a metal matrix of Al/SiCp. They have done the turning operation and studied the machinability charac-teristics of carbide cutting insert with TiN coating. Experiments were conducted in dry machining environment (Figure 3.3).

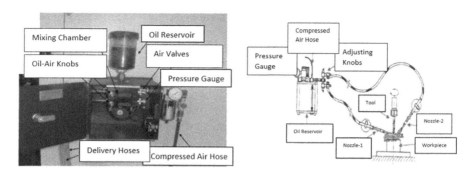

FIGURE 3.3 (a) MQL unit; (b) lubricant delivery system [21]. MQL, minimum quantity of lubricant.

After machining, continuous saw tooth chips were gathered, and these are of fragmented type. Cutting speed was found to be the most significant variable, which affects the wear, but given values of feed are observed to be the most effective parameters for surface roughness. Higher R2 value gives the highly significant regression model. After experiments, it was concluded that the predicted values are very near to experimental results. Schulze and Zanger [24] performed experiments on titanium alloy with variable cutting speed, and segmented chips were formed. The mechanical and thermal load variations were studied. Finite element model with self-developed continuous remeshing method was developed to form the serrated chips. Tool wear, stresses, temperatures, and velocity of the tool face with tool tip had been investigated. Sahoo and Pradhan [25] have taken cutting speed, feed, and depth of cut as the process parameters and found their effects on the flank wear and surface finish while turning of Al/SiCp metal using uncoated tungsten carbide inserts. The experiments were carried out under dry environment. The design of experiments was taken using L9 array of Taguchi. Premature failure of tool did not happen. The observed optimal parametric combinations are v3–f1–d3 and v1–f1–d3 for surface roughness and flank wear, respectively.

Zhang et al. [12] studied the saw tooth chip formation in experiment. They used Inconel 718 as raw material and N/TiN-coated cutting tool. They performed experiments using dry experimental conditions. The following observations are obtained: (i) at high speed cutting, saw tooth chips are formed because of formation of cyclic crack, and (ii) during turning process, they studied cutting force components keeping all parameters constant. The periodical deviation of the chip thickness results in the fluctuation of the cutting force components. Zetek et al. [26] have done optimization process to find out that what is important during optimization process. Standard finishing processes were carried out, but cutting edge radiuses were under main consideration. It is necessary to use a perfect suitable device to view a complex cutting tool and cutting process. In the experiment, the results define the selection criteria of cutting tool and tool geometry for machining of Inconel (Figure 3.4).

In experiments, they used a tool with edge-shaped cutting tip. The tool wear is affected by all of these selected parameters. For better dependability, it is necessary to get tool wear without notches or other defects. It results to the increase in tool

Machinability Aspects of Difficult-to-Cut Materials

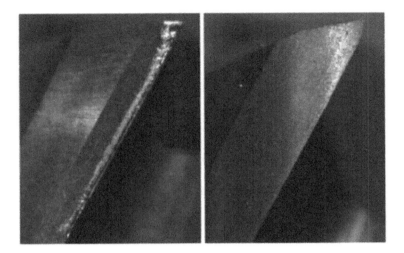

FIGURE 3.4 Cutting tool without protected facet, wear on the cutting tool [26].

efficiency and overall safety in machining of superalloys. Caliskan and Kucukkose [27] used the CN/TiAlN-coated carbide tools and studied wear behavior with cutting performance in milling operation of Ti-6Al-4V taking dry conditions during machining. The investigation is done for the surface finish with respect to cutting forces. SEM was conducted along with energy-dispersive X-ray spectroscopy to study the structural and compositional characterization of worn workpieces. Results show that the abrasive wear is the dominant tool failure on coated tools. Milling of Ti-6Al-4V was done using CN/TiAlN-coated carbide tools, and the 15% of longer lifetime was obtained with tool wear.

Zhan et al. [28] used microgrooved tool in turning operation. Their main focus was on the production and consumption of cutting energy. There was reliability of the models through sensible cutting experiment as shown in Figure 3.5. Results also show the significant effect of speed, depth taken, and applied feed on cutting energy. Parida and Maity [29] used Inconel 718 in the room as well as at elevated temperature. Analysis was done by both experiments and FEM simulation. Cutting force, chip size, shape, temperature, and thrust force were taken under observation. On comparing simulation data and experimental outputs, it was found to be partial validation. Hence, the simulation is good to predict results for those experiment conditions in which the experiments are costly, difficult, and time-consuming. It was also found that when the workpiece temperature increases, the chip thickness decreased.

Maity and Pradhan [30] presented the turning operation on titanium grade 5 alloys. Uncoated carbide inserts are used. The output parameters under observation are taken, such as speed, depth and feed given on wear of tool, surface finish, and chip reduction coefficient that were studied using Taguchi L9 orthogonal array design. Multiobjective optimization on the basis of ratio analysis (MOORA) technique has been used as an optimization tool. It is a multiobjective optimization technique that considers all the attributes along with their relative importance. The optimal cutting conditions of the cutting parameters are obtained by using MOORA

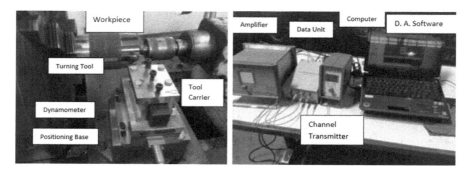

FIGURE 3.5 Cutting test machine setup [28].

coupled with Taguchi method. The used parameters are cutting speed of 112 m/min, cutting depth of 1.6 mm with given input feed of 0.04 mm/rev, and cut depth of 1.6 mm. Results declared that depth of cut gives maximum effect on surface roughness. When cutting speed increased and given feed is also increased, the wear rate of the cutting inserts increases. By increasing the given feed and cutting speed, there is decrease in chip reduction coefficient. The optical microscopic image of the chips shows that serrated chips were formed during machining of Ti-6Al-4V. Maity and Pradhan [31] performed experiments on titanium alloy (Ti-6Al-4V) using WM25CT cutting inserts. The main factors under consideration were cutting speed, depth of cut, applied feed, and their effect on cutting force, chip reduction, surface roughness, and wear of the cutting tool were analyzed. The response surface methodology (RSM) approach was used to carry out the experiments with central composite design. The obtained quadratic equations were compared with the experimental values. Chip morphology shows that side flow of chips and gap between the chips varied when process parameters are changed and also the type of chip obtained varied with change in process parameters. Damage of nose was observed when cutting speed was 160 m/min. During this, the feed was 0.14 mm/rev with depth of cut 0.75 mm.

Maity and Pradhan [32] performed experiments on Ti-6Al-4V alloy using carbide inserts by mist cooling lubrication. The considered process parameters were cutting forces, surface roughness, material removal rate, tool wear, and chip reduction. It was concluded that the best result was found when the cutting speed of tool was taken 160 m/min and feed rate was given to be 0.16 mm/rev. With these, the cutting depth of 1.6 mm was taken. Various effects were observed on chips formation due to cutting speed along with depth of cut. Long tubular and helical chips were obtained. In image analysis, as shown in Figures 3.6 and 3.7, it was observed that the free surface lamella of the chips increased when the cutting speed increases 95–160 m/min.

Pradhan and Maity [30] performed experiments on titanium alloy of grade 5 using PVD Al-Ti-N coating carbide insert (KC5010) to study the effect of the cutting variables on surface roughness. Along with this, tool wear and chip reduction coefficient were also studied. Taguchi L27 is used for design of experiments. Analysis of variance (ANOVA) is applied for the cutting variable influencing the responses. Optimal conditions were validated with confirmation test. It is declared that the

Machinability Aspects of Difficult-to-Cut Materials 59

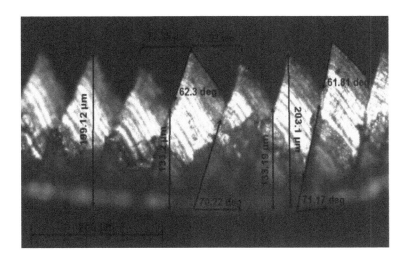

FIGURE 3.6 V-160 m/min, f-0.16 mm/rev, d-1.6 mm [32].

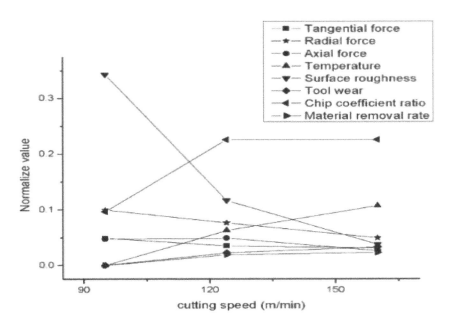

FIGURE 3.7 Effect of cutting speed on process responses [32].

optimum parameters setting has significantly improved the machining performance of titanium alloy.

Maity and Pradhan [33] explain the machining of titanium alloy grade 5. They used MT-CVD-coated cutting tool. The experiment is designed to be carried out using Taguchi L27 array layout. Three cutting variables with three levels are taken.

Desirability function analysis (DFA) approach was used to find out the optimum parameters. Herbert et al. [34] produced new cutting tool having microhole pattern. This microhole pattern controls the tribological characteristics. Various dimensions and a number of microholes pattern in various orientations were manufactured. Then a comparison study has been done between a microhole pattern insert and the normal insert. After that, the leading parameter on the machining of titanium alloy (Ti-6Al-4V) is tested. During the process, MQL method is followed for the supply of lubricant.

SEM images are taken after 10 min of machining which are shown in Figure 3.8. The adhesion and abrasion tool wear were clearly observed. If the comparison is done for the results obtained from normal insert with microtextured inserts, then results show the less flank wear. In experiments, the cutting insert with microhole-type texture designs 1 and 2 as given in picture 8 were used. Higher cutting temperature is the leading factor of judging machinability. It is apparent from the output table that in case of microhole-textured insert, the adhesion of chips is less. Along with this, the machining performance is better even at higher speed, which improves the productivity. The chips obtained from first design have extra shear bands in contrast to second design as shown in Figure 3.9.

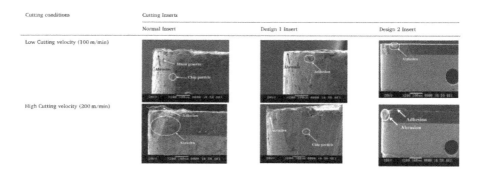

FIGURE 3.8 SEM images of tool flank wear [34].

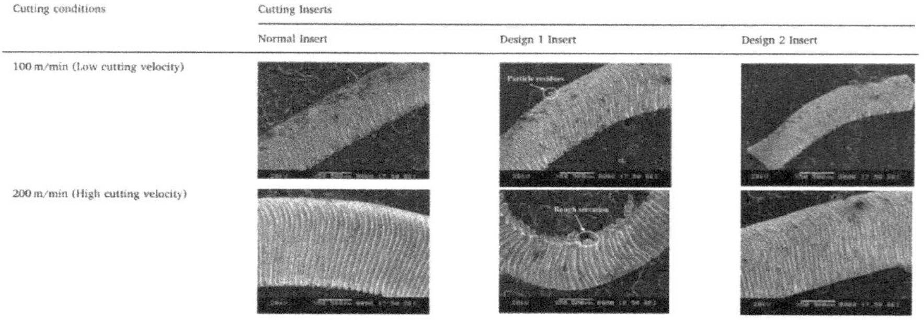

FIGURE 3.9 SEM images of the chip morphology [34].

Machinability Aspects of Difficult-to-Cut Materials

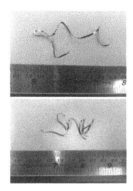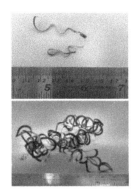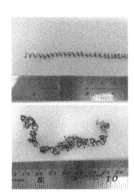

FIGURE 3.10 Measurement of produced chips [34].

It was declared that the insert with microhole reduced the friction, which results in reduction in vibration up to 30%–50%, and also the surface finish is improved to 40%, with 30% reduction in cutting temperature. Figure 3.10 shows the produced conical and helical.

For the further improvement in machinability, liquid lubricant can be used. All these parameters improve the tool life without any negative effect on performance. Machining characteristics of Ti-6Al-4V alloy were studied for the application of biomedical applications by Pradhan et al. [35–38]. It was reported that nontraditional machining with MQL technique is potential and sustainable solution for the machining industries. There is a huge potential for the machining of titanium- and magnesium-based alloy using nontraditional machining with MQL technique for biomedical application [39–57].

3.3 DISCUSSION AND FUTURE WORK

Recently, numerous nontraditional techniques are available such as MQL, coated and uncoated cutting tools, flood lubrication, and so on. But all these approaches have constraints such as unavailability of resources, requirement of special setup, problem of disposal of coolants, trained operator requirement, and so on. Therefore, the nontraditional technique such as microgrooved surface of cutting inserts is a superior technique for reduction of cutting temperature, tool wear, and cutting forces with increase of tool life. It reduces the tool chip contact area and also provides a reservoir of lubricant. But the microtexturing is also a typical task as it requires special machine for fabrication of patterns. The chip morphology also describes the rank of machinability. In near future, the research can be followed on metals such as titanium and its alloys, Inconel, nickel, and all material with low thermal expansion coefficient on which such machining improvement techniques are still to be discovered. Also work has to be done on design and simulation of several microgrooved patterns, and subsequent investigation of tool wear and chip forms for textured insert has to be carried out.

3.4 CONCLUSIONS

After comprehensive study of existing literature on textured cutting tools, the following observations have been made:

- For an efficient machining, good surface finish is an essential requirement.
- The quality of surface finish declines as the wear increases. It may be due to poor lubricating condition or lack of cooling at cutting interface.
- Machinability improves by using the textured tools for difficult-to-machine materials such as Ti- and Ni-based alloys.
- Textured tools by using solid lubricants such as MoS2 and WS2 depict an effective result in reducing surface roughness, cutting forces, cutting temperature, and coefficient of friction. Thus, it reduces the energy required for the machining and increases tool life significantly.
- The attributes for the aforementioned improvements are as follows:
  a. Contact area decreases between tool and chip
  b. Self-lubricating property at the cutting zone of tool and chips
- Effectiveness of textured tools merely depends on designs of textures, alignment with cutting edge, area of texture, and orientation.
- The design and orientation of texture according to the direction of chip flow is very significant for the enrichment in machinability of cutting tool. Because of fewer adherences in perpendicular texture orientation, it is more effective than parallel textures.

REFERENCES

1. Habrat W, Markopoulos AP, Motyka M, Sieniawski J. Machinability. In *Nanocrystalline Titanium* 2019: Elsevier. Netherland. p. 209–236.
2. Trent, E.M. and P.K. Wright, *Metal Cutting.* Boston : Butterworth-Heinemann, 2000.
3. Youssef, H.A., *Machining of Stainless Steels and Super Alloys: Traditional and Nontraditional Techniques.* Egypt: John Wiley & Sons, 2015.
4. Kus A, Isik Y, Cakir MC, Coşkun S, Özdemir K. Thermocouple and infrared sensor-based measurement of temperature distribution in metal cutting. Sensors. 2015, 15(1):1274–91.
5. Cherukuri, R. and P. Molian, Lathe turning of titanium using pulsed laser deposited, ultra-hard boride coatings of carbide inserts. *Machining Science and Technology*, 2003. **7**(1): p. 119–135.
6. Boubekri, N., V. Shaikh, and P.R. Foster, A technology enabler for green machining: minimum quantity lubrication (MQL). *Journal of Manufacturing Technology Management*, 2010. **21**(5): p. 556–566.
7. Tai, B.L., et al., Minimum quantity lubrication (MQL) in automotive powertrain machining. *Procedia CIRP*, 2014. **14**: p. 523–528.
8. Elshwain, A., N. Redzuan, and N.M. Yusof, Machinability of nickel and titanium alloys under of gas-based coolant-lubricants (cls)—a review. *International Journal of Research in Engineering and Technology*, 2013. **2**(11): p. 690–702.
9. Vasumathy, D., A. Meena, and M. Duraiselvam, Experimental study on evaluating the effect of micro textured tools in turning aisi 316 austenitic stainless steel. *Procedia Engineering*, 2017. **184**: p. 50–57.

Machinability Aspects of Difficult-to-Cut Materials

10. Biermann, D. and M. Steiner, Analysis of micro burr formation in austenitic stainless steel X5CrNi18-10. *Procedia CIRP*, 2012. **3**: p. 97–102.
11. Enomoto, T. and T. Sugihara, Improvement of anti-adhesive properties of cutting tool by nano/micro textures and its mechanism. *Procedia Engineering*, 2011. **19**: p. 100–105.
12. Zhang, S., et al., Saw-tooth chip formation and its effect on cutting force fluctuation in turning of Inconel 718. *International Journal of Precision Engineering and Manufacturing*, 2013. **14**(6): p. 957–963.
13. Gente, A., H.W. Hoffmeister, and C.J. Evans, Chip formation in machining Ti6Al4V at extremely high cutting speeds. *CIRP Annals – Manufacturing Technology*, 2001. **50**(1): p. 49–52.
14. Ribeiro, M.V., M.R.V. Moreira, and J.R. Ferreira, Optimization of titanium alloy (6Al–4V) machining. *Journal of Materials Processing Technology*, 2003. **143–144**: p. 458–463.
15. Hua, J. and R. Shivpuri, Prediction of chip morphology and segmentation during the machining of titanium alloys. *Journal of Materials Processing Technology*, 2004. **150**(1–2): p. 124–133.
16. Che-Haron, C.H. and A. Jawaid, The effect of machining on surface integrity of titanium alloy Ti–6% Al–4% V. *Journal of Materials Processing Technology*, 2005. **166**(2): p. 188–192.
17. Umbrello, D., Finite element simulation of conventional and high speed machining of Ti6Al4V alloy. *Journal of Materials Processing Technology*, 2008. **196**(1–3): p. 79–87.
18. Calamaz, M., D. Coupard, and F. Girot, A new material model for 2D numerical simulation of serrated chip formation when machining titanium alloy Ti–6Al–4V. *International Journal of Machine Tools and Manufacture*, 2008. **48**(3–4): p. 275–288.
19. Sun, S., M. Brandt, and M. Dargusch, Characteristics of cutting forces and chip formation in machining of titanium alloys. *International Journal of Machine Tools and Manufacture*, 2009. **49**(7–8): p. 561–568.
20. Obikawa, T., Y. Asano, and Y. Kamata, Computer fluid dynamics analysis for efficient spraying of oil mist in finish-turning of Inconel 718. *International Journal of Machine Tools and Manufacture*, 2009. **49**(12): p. 971–978.
21. Da Silva, R., et al., Tool wear analysis in milling of medium carbon steel with coated cemented carbide inserts using different machining lubrication/cooling systems. *Wear*, 2011. **271**(9–10): p. 2459–2465.
22. Schulze, V. and F. Zanger, Numerical analysis of the influence of Johnson-cook-material parameters on the surface integrity of Ti-6Al-4V. *Procedia Engineering*, 2011. **19**: p. 306–311.
23. Sahoo, A.K. and S. Pradhan, Modeling and optimization of Al/SiCp MMC machining using Taguchi approach. *Measurement*, 2013. **46**(9): p. 3064–3072.
24. Zanger, F. and V. Schulze, Investigations on mechanisms of tool wear in machining of Ti-6Al-4V using FEM simulation. *Procedia CIRP*, 2013. **8**: p. 158–163.
25. Sahoo, A.K., S. Pradhan, and A. Rout, Development and machinability assessment in turning Al/SiCp-metal matrix composite with multilayer coated carbide insert using Taguchi and statistical techniques. *Archives of Civil and Mechanical Engineering*, 2013. **13**(1): p. 27–35.
26. Zetek, M., I. Česáková, and V. Švarc, Increasing cutting tool life when machining Inconel 718. *Procedia Engineering*, 2014. **69**: p. 1115–1124.
27. Calıskan, H. and M. Kucukkose, The effect of aCN/TiAlN coating on tool wear, cutting force, surface finish and chip morphology in face milling of Ti6Al4V superalloy. *International Journal of Refractory Metals and Hard Materials*, 2015. **50**: p. 304–312.

28. Jiang, H., et al., Prediction and experimental research on cutting energy of a new cemented carbide coating micro groove turning tool. *The International Journal of Advanced Manufacturing Technology*, 2017. **89**(5–8): p. 2335–2343.

29. Parida, A.K. and K. Maity, Effect of nose radius on forces, and process parameters in hot machining of Inconel 718 using finite element analysis. *Engineering Science and Technology, an International Journal*, 2017. **20**(2): p. 687–693.

30. Pradhan, S. and K. Maity, Investigation of surface roughness, Tool Wear and Chip Reduction Coefficient during Machining of Titanium Alloy with PVD Al-Ti-N Coating Carbide Insert. *Analysis (DFA)*, 2016. **2**: p. 4.

31. Maity, K. and S. Pradhan, Study of chip morphology, flank wear on different machinability conditions of titanium alloy (Ti-6Al-4V) using response surface methodology approach. *International Journal of Materials Forming and Machining Processes (IJMFMP)*, 2017. **4**(1): p. 19–37.

32. Maity, K. and S. Pradhan, Study of process parameter on mist lubrication of Titanium (Grade 5) alloy. In *IOP Conference Series: Materials Science and Engineering*. 2017. IOP Publishing.

33. Maity, K. and S. Pradhan, Investigation of tool wear and surface roughness on machining of titanium alloy with MT-CVD cutting tool. In *IOP Conference Series: Materials Science and Engineering*. 2018. IOP Publishing.

34. Rao, C.M., S.S. Rao, and M.A. Herbert, Development of novel cutting tool with a microhole pattern on PCD insert in machining of titanium alloy. *Journal of Manufacturing Processes*, 2018. **36**: p. 93–103.

35. Pradhan, S., et al., Investigation of machining characteristics of hard-to-machine Ti-6Al-4V-ELI alloy for biomedical applications. *Journal of Materials Research and Technology*, 2019. **8**(5): p. 4849–4862.

36. Prakash, C., et al., *Biomanufacturing*. Switzerland: Springer nature, 2019.

37. Singh, H., S. Singh, and C. Prakash, Current trends in biomaterials and biomanufacturing, in *Biomanufacturing*. Switzerland: Springer nature, 2019.

38. Pradhan, S., et al., Micro-machining performance assessment of Ti-based biomedical alloy: a finite element case study, in *Biomanufacturing*. Switzerland: Springer nature, 2019. p. 157–183.

39. Prakash, C., et al., Electric discharge machining–a potential choice for surface modification of metallic implants for orthopedic applications: a review. *Proceedings of the Institution of Mechanical Engineers, Part B: Journal of Engineering Manufacture*, 2016. **230**(2): p. 331–353.

40. Prakash, C., et al., Experimental investigations in powder mixed electric discharge machining of Ti–35Nb–7Ta–5Zrβ-titanium alloy. *Materials and Manufacturing Processes*, 2017. **32**(3): p. 274–285.

41. Prakash, C., et al., Processing and characterization of novel biomimetic nanoporous bioceramic surface on β-Ti implant by powder mixed electric discharge machining. *Journal of Materials Engineering and Performance*, 2015. **24**(9): p. 3622–3633.

42. Prakash, C., et al., Multi-objective optimization of powder mixed electric discharge machining parameters for fabrication of biocompatible layer on β-Ti alloy using NSGA-II coupled with Taguchi based response surface methodology. *Journal of Mechanical Science and Technology*, 2016. **30**(9): p. 4195–4204.

43. Prakash, C. and M. Uddin, Surface modification of β-phase Ti implant by hydroaxyapatite mixed electric discharge machining to enhance the corrosion resistance and in-vitro bioactivity. *Surface and Coatings Technology*, 2017. **326**: p. 134–145.

44. Prakash, C., et al., Powder mixed electric discharge machining: an innovative surface modification technique to enhance fatigue performance and bioactivity of β-Ti implant for orthopedics application. *Journal of Computing and Information Science in Engineering*, 2016. **16**(4): p. 041006, 1–9

45. Prakash, C., et al., Synthesis, characterization, corrosion and bioactivity investigation of nano-HA coating deposited on biodegradable Mg-Zn-Mn alloy. *Surface and Coatings Technology*, 2018. **346**: p. 9–18.
46. Prakash, C., et al., Multi-objective particle swarm optimization of EDM parameters to deposit HA-coating on biodegradable Mg-alloy. *Vacuum*, 2018. **158**: p. 180–190.
47. Prakash, C., et al., Synthesis and characterization of Mg-Zn-Mn-HA composite by spark plasma sintering process for orthopedic applications. *Vacuum*, 2018. **155**: p. 578–584.
48. Prakash, C., et al., Surface modification of Ti-6Al-4V alloy by electrical discharge coating process using partially sintered Ti-Nb electrode. *Materials*, 2019. **12**(7): p. 1006.
49. Prakash, C., et al., On the influence of nanoporous layer fabricated by PMEDM on β-Ti implant: biological and computational evaluation of bone-implant interface. *Materials Today: Proceedings*, 2017. **4**(2): p. 2298–2307.
50. Prakash, C., et al., Bio-inspired low elastic biodegradable Mg-Zn-Mn-Si-HA alloy fabricated by spark plasma sintering. *Materials and Manufacturing Processes*, 2019. **34**(4): p. 357–368.
51. Prakash, C., et al., Effect of surface nano-porosities fabricated by powder mixed electric discharge machining on bone-implant interface: an experimental and finite element study. *Nanoscience and Nanotechnology Letters*, 2016. **8**(10): p. 815–826.
52. Gupta, M.K., et al., Machinability investigations of Inconel-800 super alloy under sustainable cooling conditions. *Materials*, 2018. **11**(11): p. 2088.
53. Pramanik, A., A. Basak, and C. Prakash, Understanding the wire electrical discharge machining of Ti6Al4V alloy. *Heliyon*, 2019. **5**(4): p. e01473.
54. Pramanik, A., et al., Optimizing dimensional accuracy of titanium alloy features produced by wire electrical discharge machining. *Materials and Manufacturing Processes*, 2019. **34**(10): p. 1083–1090.
55. Prakash, C., et al., Fabrication of low elastic modulus Ti50Nb30HA20 alloy by rapid microwave sintering technique for biomedical applications. *Materials Today: Proceedings*, 2020. **21**: p. 1713–1716.
56. Prakash, C., et al., Microwave sintering of porous Ti–Nb-HA composite with high strength and enhanced bioactivity for implant applications. *Journal of Alloys and Compounds*, 2020: p. 153774.
57. Prakash, C. and S. Singh, On the characterization of functionally graded biomaterial primed through a novel plaster mold casting process. *Materials Science and Engineering: C*, 2020. 110: p. 110654.

4 Micromachining

Venkatasreenivasula Reddy Perla
and Rathanraj K.J.
Department of Industrial Engineering and Management,
B.M.S. College of Engineering, Bengaluru, India.

CONTENTS

4.1 Introduction 68
4.2 Conventional Micromachining 69
4.3 Nonconventional Micromachining 71
    4.3.1 Ultrasonic Micromachining..71
        4.3.1.1 Working Principle 71
        4.3.1.2 Tool Material...71
        4.3.1.3 Tool Feed Mechanism..72
        4.3.1.4 Abrasive Slurry System 72
        4.3.1.5 Oscillating System 72
        4.3.1.6 Process Parameters 73
        4.3.1.7 Effect of Process Parameters on Material Removal Rate....73
        4.3.1.8 Advantages..74
        4.3.1.9 Limitations..74
        4.3.1.10 Applications 74
    4.3.2 Abrasive Jet Micromachining..74
        4.3.2.1 Working Principle 75
        4.3.2.2 Process Parameters 75
        4.3.2.3 Abrasive Material..76
        4.3.2.4 Gas Medium..77
        4.3.2.5 Nozzle 77
        4.3.2.6 Effect of Material Removal Rate 77
        4.3.2.7 Advantages..77
        4.3.2.8 Limitations..77
        4.3.2.9 Applications 78
    4.3.3 Electrochemical Micromachining 78
        4.3.3.1 Working Principle 78
        4.3.3.2 Process Parameters 79
        4.3.3.3 General Material Removal Rate Model for
                Electrochemical Micromachining 81
        4.3.3.4 Advantages..82
        4.3.3.5 Limitations..82
        4.3.3.6 Applications 82

4.3.4	Electrodischarge Micromachining	83
	4.3.4.1 Working Principle	83
	4.3.4.2 Components of Electrodischarge Micromachining	84
	4.3.4.3 Process Parameters	85
	4.3.4.4 Variants of Electrodischarge Micromachining	85
	4.3.4.5 Advantages	86
	4.3.4.6 Limitations	87
	4.3.4.7 Applications	87
4.3.5	Laser Beam Micromachining	88
	4.3.5.1 Working Principle	88
	4.3.5.2 Mechanism of Material Removal	89
	4.3.5.3 Laser Mask Projection Technique	91
	4.3.5.4 Effect of Laser Beam Intensity	91
	4.3.5.5 Material Removal Rate in Pulsed Laser	92
	4.3.5.6 Advantages	94
	4.3.5.7 Limitations	94
	4.3.5.8 Applications	94
4.3.6	Electron Beam Micromachining	95
	4.3.6.1 Working Principle	95
	4.3.6.2 Electron Beam Micromachining Equipment	96
	4.3.6.3 Components of Electron Beam Micromachining	96
	4.3.6.4 Process Parameters	98
	4.3.6.5 Advantages	99
	4.3.6.6 Limitations	99
	4.3.6.7 Applications	100
4.3.7	Plasma Arc Micromachining	100
	4.3.7.1 Working Principle	101
	4.3.7.2 Process Parameter	101
	4.3.7.3 Advantages	102
	4.3.7.4 Limitations	102
	4.3.7.5 Applications	102
4.4	Conclusions and Future Study	102
References		105

4.1 INTRODUCTION

In the present manufacturing era, the demand for high-accuracy components in various fields such as biomedical, production, meteorological, aerospace, electronics, automobile, and so on has been increasing extensively. Especially in the fields of medical, aerospace, and MEMS (Micro-Electro-Mechanical Systems), the applications of micromachining are excessively expanded. Advances in material science increased the innovation of new materials with rich mechanical properties, leading to upgradation of machining processes. Micromachining processes can be mainly classified as two types: conventional and nonconventional processes; these are further classified based on several factors as shown in Figure 4.1.

Micromachining

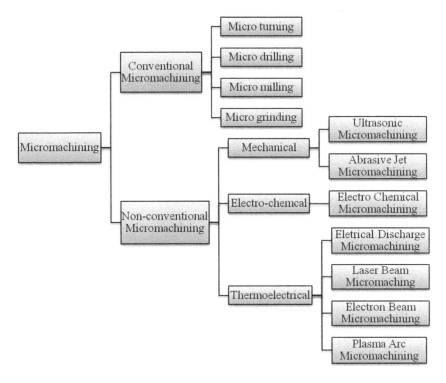

FIGURE 4.1 Classification of micromachining.

4.2 CONVENTIONAL MICROMACHINING

Machining in conventional micromachining processes is based on frictional shear between tool and workpiece; these processes are miniaturization of existing conventional methods for microlevel applications. The different processes are microturning, microdrilling, micromilling, microgrinding, and so on. In all these processes, both tool and workpiece are in direct mechanical contact, and the machining zones are connected to different sensors to measure cutting forces and temperatures and to observe changes in operational parameters (as shown in Figures 4.2 and 4.3) [13]. The accuracies of the machined surface depend on several parameters such as process variables, effects of vibration, working materials, contact variables, and so on. These processes have higher material removal rates (MRRs) when compared with nonconventional machining [27]. The primary requirement of conventional processes is that the hardness of tool material must be greater than the workpiece material because the cutting principle is mechanical wear. To machine the workpiece material, the applied stress should be more than ultimate tensile strength of the material. [5] To initiate shearing on the workpiece surface with tiny tools, very high spindle speeds (about 60,000 to 4,00,000 rpm) are required [29]. Most of the microdrills are prepared by using microgrinding process [92]. To achieve accurate dimensional tolerances in these conventional processes, instructions should be given

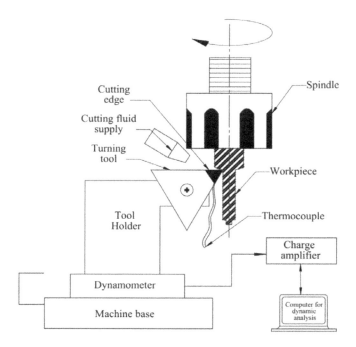

FIGURE 4.2 Microturning.

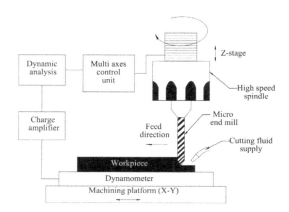

FIGURE 4.3 Micromilling.

with a computer numerical controller (CNC). Today's micromachining equipment is commercially available with CNC controllers and desired sensors to enhance the machining accuracy. The processes working based on mechanical energy for material removal workpiece need not be electrically conductive.

Problems with conventional micromachining such as excessive tool wear, condition of the tool hardness, high vibrational disturbances, poor machining accuracy, lack of variety in the process, difficulties in complex shape machining, larger cutting

Micromachining 71

forces, excessive heat generation due to high revolution speeds, burr formation, processing material restriction, and so on motivated the manufacturing field toward generation of new machining methods (nonconventional machining methods) to overcome these difficulties [3,20,23,44,45,50,80,93].

4.3 NONCONVENTIONAL MICROMACHINING

The collection of machining processes that utilize the energy source (mechanical, electrical, chemical, thermal energy, or combination of these energies) for material removal of complicated materials are named non-conventional machining methods. These processes are also called as advanced micromachining methods and nontraditional machining methods. In these processes, there is no direct physical contact between workpiece and tool. Materials such as highly brittle, high tempered steels, superalloys, metal carbides, and high-performance ceramics can be processed comfortably. [17]

4.3.1 Ultrasonic Micromachining

Ultrasonic micromachining (USMM) is a type of nonconventional machining method most widely employed to machine very hard and brittle materials by using mechanical action between abrasive and the workpiece surface. The material removal in USMM is due to the hammering effect. Brittle materials give better results compared with ductile materials [30]. So, it has been applied successfully to ceramics, ferrites, carbides, glasses, cermets, some superalloys, and so on.

4.3.1.1 Working Principle

USMM is a type of grinding process to abrade the workpiece with abrasive particles. In this, the continuous abrasive slurry is pumped between tool and workpiece, and high-frequency, low-amplitude vibrations are applied to the tool. When the tool is vibrating with high frequency, the impact loads produced from the tool will be transmitted to abrasive particles against workpiece, which in turn produces impact loads to the workpiece. When the impact loads are acting on the workpiece, if the workpiece material is made of a brittle material, because of lower toughness, the workpiece will get fracturing due to "brittle fracture." Hence, the mechanism of material removal is brittle fracture when machining brittle materials and plastic deformation when machining ductile materials. The impact loads of hard abrasive grains produce minute chips by wearing workpiece; these chips will be carried away by the slurry [74,89].

4.3.1.2 Tool Material

When the impact loads act on the abrasive particles, equal and opposite force produced by the abrasives will be acting onto the tool. Therefore, if the tool is made of a hard and brittle material, the brittle fracturing will take place on the tool also. Hence the tool wear will be higher. To minimize the tool wear, the tool must be made of a soft material. Generally used tool materials are mild steel, copper, brass, and so on. If the tool material is soft and ductile, it is easy to reproduce the tool to its initial

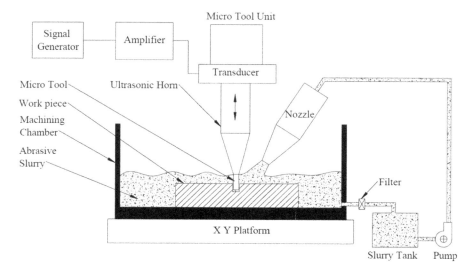

FIGURE 4.4 Working principle and experimental setup of USMM. USMM, ultrasonic micromachining.

shape after machining. To produce accurate results, tool must be prepared with smaller dimensions depending on the abrasive grain size and pumping pressure [38].

4.3.1.3 Tool Feed Mechanism

The feed mechanism is used to bring the tool slowly toward the workpiece and to provide the required amount of cutting force to withstand the forces of machining and then to stop the cutting forces when required machining depth has been obtained and also to bring the tool to starting position. The various types of mechanisms used are (i) gravity feed mechanism, (ii) spring-loaded feed mechanism, (iii) pneumatic or hydraulic feed mechanism, and (iv) motor-controlled feed mechanism.

4.3.1.4 Abrasive Slurry System

The main function of abrasive slurry is to avoid direct contact between tool and workpiece by flowing continuously between them. The common abrasive materials can be aluminm oxide, boron carbide, and silicon carbide grains in 3–10 μm range. Generally, boron carbide is used with water as slurry, and silicon carbide is used with paraffin as slurry. The other commonly used abrasives are diamond dust and boron silicon carbide. The ratio of abrasive and liquid in abrasive slurry varies from 1:4 to 1:1 by weight. The abrasive slurry should be fed continuously to avoid drying up at the tool face.

4.3.1.5 Oscillating System

Oscillating system consists of an ultrasonic transducer, ultrasonic horn, and a coupler. The ultrasonic transducer is a major component of the oscillating system setup. This transducer produces the vibrations of frequency above 20 kHz and amplitude

Micromachining

of 0.8–5 μm for USMM. Ultrasonic horn's duty is to transmit vibrations from transducer to tool, and it also amplifies the vibration amplitudes. The nonmachining end of the tool is fixed to the horn by silver brazing to increase vibrational efficiency instead of mechanical clamping. The coupler may be used for clamping of the transducer and horn assembly. The horn can be used with different shapes such as tapered or conical, exponential, stepped, and so on.

4.3.1.6 Process Parameters

The process parameters of USMM are categorized into workpiece and tool-related, abrasive parameters, and vibrational and horn parameters. Each variable for the respective parameter is listed in Figure 4.5. Researchers vary the parameters to find the best optimal conditions for output response variables. The performance variables include MRR, surface roughness, tool wear, hole overcut size, circularity, cylindricity, and so on.

4.3.1.7 Effect of Process Parameters on Material Removal Rate (Figure 4.6)

1. **Frequency** (F): Increase in frequency increases MRR slightly because the number of oscillations of the tool in a given time will increase.
2. **Amplitude** (a_0): By increasing the amplitude, the MRR will increase since the horn will vibrate in longer lengths to increase vibrational amplitude, and the higher impact loads fall on the workpiece to increase MRR.
3. **Abrasive grain size** (d_g): With increase in grain size, there will be an increase in the size of crater generated because the larger grains cause more material removal. Hence, MRR will increase. Surface roughness will be poor.
4. **Feed force** (f): Feed force is directly changing the value of MRR since the impact load causing material removal will increase directly. The higher the feed force, the higher the MRR values and the lower the surface finish [43].

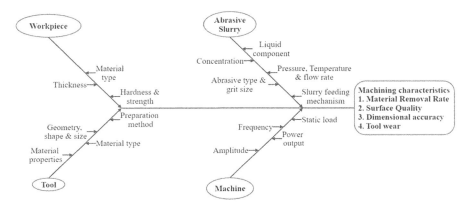

FIGURE 4.5 Cause and effect diagram of USMM characteristics [31]. USMM, ultrasonic micromachining.

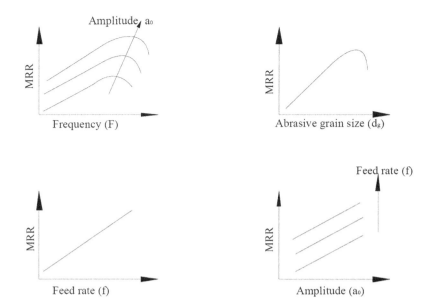

FIGURE 4.6 Effect of process variables on MRR. MRR, material removal rate.

4.3.1.8 Advantages
1. Highly brittle materials can be machined very easily.
2. Smaller noncircular holes can be produced.
3. Workpiece needs not be electrically conductive.
4. Machined surface is free from thermal effects and residual stresses.

4.3.1.9 Limitations
1. MRR is very low.
2. Tool wear is high.
3. For deeper holes, the gap present between workpiece and hole will be less.

4.3.1.10 Applications (Figure 4.7)
1. This method can produce holes as small as 20 μm as diameter.
2. Dentists use this method to produce holes in human teeth because it is a painless drilling method.
3. This method is used for drilling, grinding, profiling, and coining on materials such as stainless steel, glass, ceramic, carbide, quartz, semiconductors, and so on.

4.3.2 Abrasive Jet Micromachining

Abrasive jet micromachining (AJMM) method is the process in which the material is removed from the workpiece surface using the mechanical impact energy of high pressure using abrasive material passing through a nozzle [1].

Micromachining

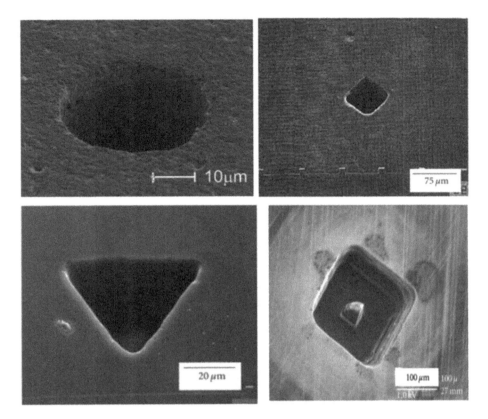

FIGURE 4.7 Microhole in Al_2O_3 machined with USMM, square hole and triangular hole in silicon, 3D cavity by USMM [15]. (Copied with permission.) USMM, ultrasonic micromachining.

4.3.2.1 Working Principle

In AJMM, very high air pressure is supplied from the compressor to develop kinetic energy for the fine abrasive stream. Air and abrasive mix pass through the nozzle with very high velocities (500–600 m/s) to create a microfracture on the surface of the workpiece. When abrasive creates microcraters, microchips will be produced. Both microchips and abrasive particles will be taken away by compressed gas or air stream to allow the fresh surface to come in action to expose the abrasive stream as shown in Figure 4.8. Nozzle tip distance is the distance between the nozzle tip and workpiece, which is in the order of microns. Moreover, while processing brittle materials mask projections were applied [2,8,24,40,52,72].

4.3.2.2 Process Parameters [68]

The process parameters of AJMM are categorized and shown in Figure 4.9 according to the type of group it belongs to.

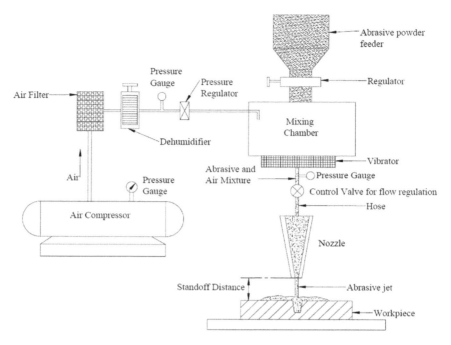

FIGURE 4.8 Abrasive jet micromachining.

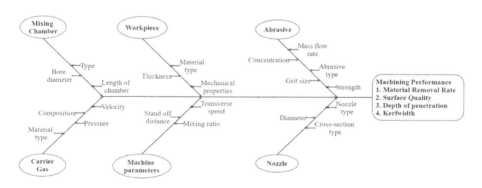

FIGURE 4.9 Cause and effect diagram for AJMM performance. AJMM, abrasive jet micromachining.

4.3.2.3 Abrasive Material

The commonly used abrasive materials in AJMM are silicon carbide, aluminum oxide, glass powder, and $CaMgCO_3$ (dolomite). The machining efficiency is based on size, shape, hardness, and strength of the abrasive particles. For cleaning and etching process, $CaMgCO_3$ (dolomite) is used as abrasive. Similarly, SiC is used for very hard materials, Al_2O_3 is used for general purpose, and glass powder is used as abrasive material for polishing purpose [25,26,28].

Micromachining

4.3.2.4 Gas Medium [35]

Commonly used gas medium is air, which must be filtered to remove moisture content and other contaminants. Other alternatives for air are carbon dioxide or nitrogen. The gas medium should be nontoxic and cheap. Oxygen usage is restricted due to violent chemical actions with abrasive particles or microchips of the workpiece.

4.3.2.5 Nozzle [85]

Nozzle should be designed to withstand high abrasion wear of the jet stream. Nozzle is made by using hard materials such as tungsten carbide (WC) and sapphire. Sapphire is the most suitable nozzle material, since the life of sapphire is 10 times the life of WC.

4.3.2.6 Effect of Material Removal Rate

MRR of AJMM depends on several factors such as abrasive flow rate, abrasive mixing ratio, nozzle diameter, workpiece material, the velocity of the jet, hardness and size of abrasive particles, air pressure, and nozzle tip distance [10,36].

$$\text{Mixing ratio} = \frac{\text{mass flow rate of abrasive particles}}{\text{mass flow rate of air}}$$

By increasing abrasive mixing ratio, the MRR will increase up to certain limit because the number of particles hitting workpiece surface will increase first, and further increase causes no room for gas medium to remove microchips, resulting poor removal rate. By increasing jet velocity kinetic energy, the impact will increase and cause more MRR. When we increase the nozzle tip distance or standoff distance, the MRR increases first and reaches optimum value and then decreases gradually. This is due to a small increase in standoff distance. The kinetic energy will increase with increase in velocity. When further increased, atmospheric drag will come into play, decreasing the velocity of the jet and causing low MRR values (as shown in Figure 4.10).

4.3.2.7 Advantages
1. Hard and brittle materials can be machined easily [79].
2. There are no thermal stresses on the finished surface.
3. Capital and operating cost is low, and power consumption is low.
4. Microfeatures with 100 µm can be produced up to small thickness.

4.3.2.8 Limitations [49]
1. Machining of ductile materials gives low MRR values.
2. This method produces loud noise and dusty environment.
3. Nozzle gets damaged with short nozzle tip distance
4. Abrasive reuse is not possible.

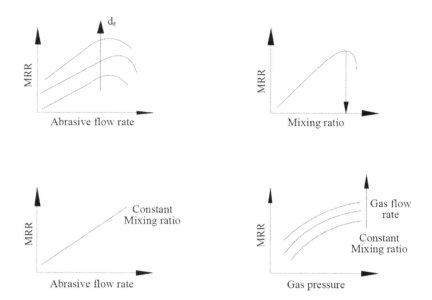

FIGURE 4.10 Effect of AJMM process variables on MRR. AJMM, abrasive jet micromachining; MRR, material removal rate.

4.3.2.9 Applications
1. By replacing gas medium by water medium, abrasive water jet machine (AWJM) is developed for larger working scales.
2. AJMM is used for deburring and polishing of plastic components [11,76].
3. This method is used for removing glue and paint from paintings and lather objects.
4. This method is used for etching or marking on glass and ceramic materials.

4.3.3 Electrochemical Micromachining

Electrochemical micromachining (ECMM) is a nonconventional machining process that uses both electrical and chemical energy to remove material from an electrically conductive workpiece by a constant supply of electrolyte. ECMM is an anodic dissolution process in which workpiece is the anode, tool is the cathode, and the electrolyte is pumped through the gap between the tool cathode and workpiece anode. Main components of ECMM are the power supply unit, electrodes, and electrolyte flow system. ECMM is commonly employed as post processing technique to reduce EDMM and LBMM's surface damage [67,77,78].

4.3.3.1 Working Principle
As a potential difference is applied between the electrodes, the principle of electrolysis comes into action because of which certain possible chemical reactions at the electrodes occurred. Electrolyte which is a salt solution dissolve into independent ions and water dissolve into H^+ and OH^- ions. At the anode, metal dissolute into free ions by giving electrons. These electrons are attracted by free H^+ ions to form

Micromachining

hydrogen gas which is liberated at the cathode in the form of hydrogen gas bubbles (as shown in Figure 4.11). OH⁻ ions react with metal ions to form metal hydroxides, which are left as precipitates. Along with these outcomes, small amount of metal chlorides are also evolved by the reaction between free ions of metal and chlorine, and when compared with metal hydroxides, these products are negligible. All major chemical reactions formed during electrolysis are listed in the following. This results in removal of material from the workpiece as a replica of the tool cathode. Machining efficiency of the ECMM depends on anodic behavior of the workpiece material with the electrolyte.

Reactions during machining,

$$M \rightarrow M^+ + e^-$$
$$H_2O \rightarrow H^+ + OH^-$$
$$NaCl \rightarrow Na^+ + OH^-$$
$$H^+ + e^- \rightarrow H_2 \uparrow$$
$$M^+ + OH^- \rightarrow M(OH) \downarrow$$
$$Na^+ + OH^- \rightarrow NaOH \downarrow$$

where M is workpiece material which is assumed as a metal.

4.3.3.2 Process Parameters (Figure 4.13)

1. **Electrolyte**: Electrolyte is a common salt solution (mostly NaCl and NaNO₃) with high electrical conductivity which is chemically active to remove material from the workpiece. The electron transfer from anode to cathode depends on the electrical properties of electrolyte in the interelectrode gap (IEG). Electrolyte also carries both the heat generated and reaction precipitates during the electrolysis. When we pump the electrolyte to the interelectrode gap, higher flow rates affect tool stability and cause tool eros The MRR of ECMM process is ion as the tool is micron in size. Components of electrolyte flow system were shown in Figure 4.12.

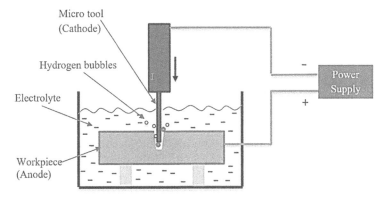

FIGURE 4.11 Working principle of ECMM. ECMM, electrochemical micromachining.

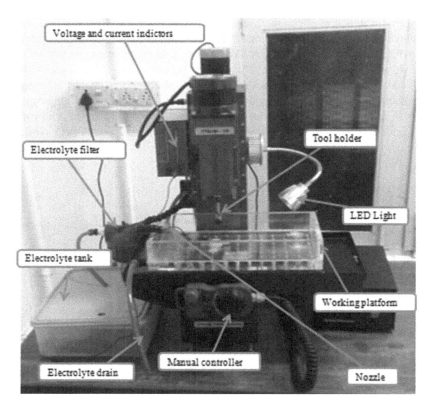

FIGURE 4.12 ECMM experimental setup [55]. (Copied with permission.) ECMM, electrochemical micromachining.

2. **Current and voltage**: The MRR of ECMM process is directly proportional to applied current, increase in current beyond optimum level leads more dimensional deviation and improper accuracy of the machined surface. The DC power supply gives a continuous voltage supply to the pulse generator to develop a pulsed voltage with a pulse on time and off time.
3. **Interelectrode gap**: An IEG is defined as the gap present between toll and workpiece, and it is in the range of 5–50 µm. This parameter is constant throughout the process.
4. **Tool design**: The tool size in ECMM is at microlevel; hence, the preparation of small tools is difficult. Since there is no tool wear, extremely hard materials are not required for preparing the tool, and it should withstand electrolyte hydraulic force at the IEG. Tool shape should be designed based on the requirement of the final shape of the surface to be desired.
5. **Material removal rate**: The MRRs of ECMM mainly depends on the tool feed rates and applied current. If there is a change in feed rate, either increase or decrease directly causes a change in current values, which influence MRRs. So, MRR is directly proportional to both tool feed rate and current [37].

Micromachining

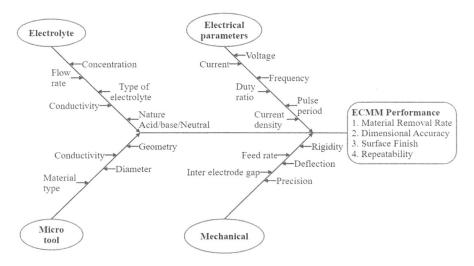

FIGURE 4.13 Cause and effect diagram of ECMM parameters. ECMM, electrochemical micromachining.

6. **Surface finish**: In ECMM, ion displacement is the basic principle of material removal. That is why, the surface finish produced by the ECMM is excellent and around 0.05–0.4 μm. The major parameters affecting the surface finish are pulse on time, pulse off time, electrolyte composition, feed rate, and so on (Figure 4.13).

4.3.3.3 General Material Removal Rate Model for Electrochemical Micromachining

ECMM is working based on Faraday's law of electrolysis. According to Faraday's first law of electrolysis,

$$\text{Mass of ions liberated} \quad M = Z \times I \times t \tag{4.1}$$

where,
Z = constant (electrochemical equivalent),
I = current flowing in Amps,
t = time in s.

Z is also equal to the number of ions liberated by the substance with 1 A passage current for 1 s through electrolytic action.

According to Faraday's second law,

$$z = \frac{1}{F} \times \frac{a_{wt}}{v} \tag{4.2}$$

From Equations 4.1 and 4.2,

$$M = \frac{1}{F} \times \frac{a_{wt}}{v} \times It \tag{4.3}$$

Therefore,

$$\text{MRR} = \frac{M}{\rho\, at} \tag{4.4}$$

$$= \frac{1 \times a_{wt} \times I}{F \times \rho\, A} \tag{4.5}$$

$$= \frac{I a_{wt}}{\rho\, AFv} \tag{4.6}$$

where,
MRR = material removal rate (cm³/s),
F = Faraday's constant = 96,500 coulombs (amp/ s),
v = valency of metal dissolved,
a_{wt} = atomic weight of material in gms A = machined area, cm²,
ρ = density of workpiece, gm/cm³ t = time in s.
Expression for current density is

$$\text{current density} = \frac{VK}{y} = \frac{\rho \times F}{Z}\,(\text{Amp/mm}^2) \tag{4.7}$$

where,
V = applied voltage (volts),
y = gap between tool and work (mm),
K = conductivity of electrolyte (mho/mm),
f = tool feed rate (mm/s).

4.3.3.4 Advantages

1. Workpiece harder than tool can be machined easily.
2. No tool wear, the same tool can be used for an infinite number of components.
3. The surface finish of the machined surface is excellent.
4. There are no thermal or mechanical stresses in workpiece after machining.
5. There is no heat-affected zone (HAZ).

4.3.3.5 Limitations

1. Energy and power consumption are high.
2. Producing sharp corners is difficult.
3. The workpiece must be electrically conductive.

4.3.3.6 Applications (Figures 4.14 and 4.15) [12,37]

1. ECMM is used for producing microholes for aerospace applications.
2. This method is used for carbide tool inserts.
3. This method is used for complex microprofiles generation.
4. This method is used for micromolds preparation.
5. This method is used for spray nozzle holes.

Micromachining

FIGURE 4.14 Micro-3D profiles machined with ECMM [6]. (Copied with permission.) ECMM, electrochemical micromachining.

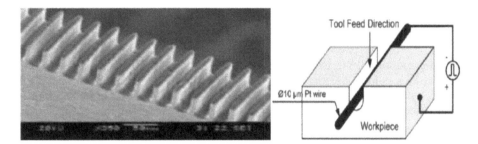

FIGURE 4.15 Microgrooves machined by wire ECMM with 10-μm-diameter platinum wire. [6]. (Copied with permission.) ECMM, electrochemical micromachining.

4.3.4 Electrodischarge Micromachining

Electrodischarge micromachining (EDMM) is a thermoelectric process used for processing of electrically conductive materials which are difficult to process using conventional techniques. It is a downscaled version of normal electrical discharge machining (EDM). In this machining process, the desired shape is machined by using a series of electric discharges (sparks) between tool and workpiece with dielectric fluid supply. The history of EDM was first observed when spark erosive effect of electrical discharge was observed in the 1770s by an English physicist Joseph Priestly. First commercial EDM was invented in 1943; after continuous development in the process with technology upgradations, efficient machines were developed (see Table 4.1).

4.3.4.1 Working Principle

EDM works based on the phenomenon of spark erosion between two electrically conductive materials, tool and workpiece. The process takes place in a nonconductive liquid medium (dielectric medium). An optimum gap between tool and workpiece is maintained constant throughout the process called IEG. The gap between electrodes varies in the range from 10 to 100 μm.

When pulsed DC power is supplied to two electrodes, and as the distance between electrodes reaches IEG, a strong electrical field is generated. As the pulse begins, the applied voltage attains breakdown voltage value, and the workpiece surface temperature will increase. Hence, the dielectric fluid present between the gap gets vaporized, and the plasma channel will be generated. Now, the plasma channel produces an

TABLE 4.1
History of EDM

Year	Event
1770	English physicist Joseph Priestley has first noted the erosive effect of electrical discharge.
1943	Russian scientists, B.R. Lazarenko, and N. I. Lazarenko invented the EDM machine based on spark erosion principle to machine hard materials such as tungsten.
1960	Wire cut EDM was invented.
1967	NC-assisted wire EDM machine was built.
1976	CNC-assisted EDM was introduced.

CNC, computer numerical controller; EDM, electrical discharge machining

instantaneous temperature range of more than 10,000°C. The sudden increase in temperature causes local heating of electrodes (more on the anode), and the workpiece material will easily melt and evaporate because the melting point of the workpiece is very much less than spark temperature. This creates a crater on the workpiece and leaves some debris on the surface [51]. These particles are flushed away by dielectric fluid flow during pulse off time. Now the pulse duration ends, a new pulse will start to create new local erosion, and the cycle continues till the required machining shape is produced. The trends of voltage and current during the machining phase for one complete pulse are related to various phases such as ignition, plasma formation, discharge phase, and ejection phase as shown in the figure 4.16.

4.3.4.2 Components of Electrodischarge Micromachining

The EDMM system contains various subsystems as shown in the figure 4.17 such as dielectric flow system, servo system, positioning system, and control unit.

1. **Pulse generation**: For the electron beam micromachining (EBMM), pulsed DC power supply is required, and the pulse generator generates pulses as per requirement. Various types of pulse generators are resistance–capacitance-type (RC-type) relaxation generator, rotary impulse-type generator, electronic pulse generator, and so on.

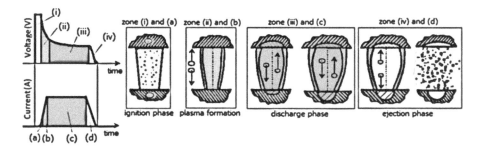

FIGURE 4.16 Voltage and current trends mapped to spark erosion mechanism phases [34].

Micromachining

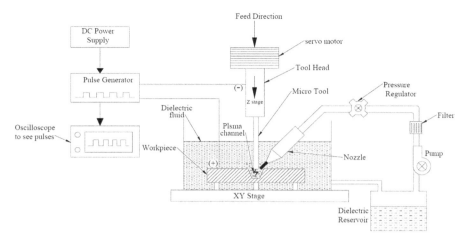

FIGURE 4.17 Schematic diagram of EDMM. EDMM, electrodischarge micromachining.

2. **Dielectric flow system**: A dielectric reservoir is mounted to circulate the fluid to machining chamber through a pump. A nozzle is employed to give sufficient fluid flow over IEG (should be optimum for higher flow velocities that tool will deflect). The main functions of dielectric fluids are (i) to insulate the workpiece from the electrode; (ii) to ionize quickly so that next discharge will be made; (iii) to cool the electrodes; and (iv) to remove debris from the machining surface.
3. **Servo control system**: The microtool movement and feed control unit of the system are monitored by this subsystem.
4. **Positioning system**: IEG for complete machining kept constant by this system. The movement of the worktable in X and Y directions and tool in Z direction are used to generate complex profiles.

4.3.4.3 Process Parameters

Parameters of EDMM are categorized into electrical, nonelectrical, electrode, dielectric, and output parameters. Every parameter of each category is shown in Figure 4.18.

4.3.4.4 Variants of Electrodischarge Micromachining

1. Micro wire-EDM

   In this process, tool material is a microwire continuously fed between two guide rollers to cut complex profiles. Since there is no chip formation and wastage, it is a material-saving process. Metal sheets and slabs with greater thickness can be cut into any profile by giving CNC assistance to the working platform. 0.1 mm diameter conductive metallic wires are used as cathode material, and most commonly used wire materials are brass, zinc-coated brass, copper, and molybdenum wires.

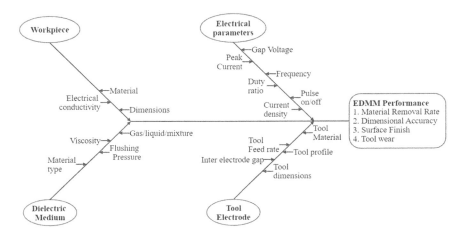

FIGURE 4.18 Process parameters of EDMM in the fishbone diagram. EDMM, electrodischarge micromachining.

2. Micro–electrodischarge grinding

    Micro–electrodischarge grinding (µEDG) is used to prepare microtools with a stationary block electrode by considering the cylindrical component of required material. Also, microwire EDG process is developed to reduce the tool diameters in tens of microns.

3. Reverse micro-EDM

    This is the most popular technique to prepare cylindrical shape of micro-end mills and microtools for various micromachining processes. In reverse micro-EDM process, the polarity between tool and workpiece is reversed to obtain more material removal on the tool surface [42].

4. Powder-mixed micro-EDM

    Most of the literatures on surface integrity of EDM surface suggest that powder-mixed EDM gives better quality [4]. Chander Prakash et al. performed several experiments on titanium alloy implants using micro-EDM and powder-mixed EDM to enhance surface quality and the applications of titanium alloys [57–66]. Furthermore, ECDM (Electro chemical discharge machining) variety is developed to minimize surface defects [32,67].

4.3.4.5 Advantages

The major advantages of micro-EDM process are as follows:

1. Any electrically conductive material irrespective of hardness, toughness, strength, and microstructure is machined easily.
2. Machined surface has good surface finish and accuracy.
3. Complex shapes can be produced easily.
4. Since there is no physical contact between tool and workpiece, machining forces are not developed during machining.
5. No chip formation and the mechanical properties of workpiece material will not affect the MRR.

Micromachining

4.3.4.6 Limitations
1. Workpiece and tool materials must be electrically conductive.
2. It is difficult to produce perfect square corners.
3. Machined surface experiences thermal effects such as recast layer, HAZ, and so on.
4. Tool wear is more.

4.3.4.7 Applications (Figures 4.19 and 4.20)
1. Micro-EDM process can successfully produce micropins, micronozzles, and microcavities on very hard components such as superalloys, tool steels, titanium alloys, and so on.
2. This process is used for machining small holes, orifices, slots in fuel injection nozzles, airbrake valves, aircraft engines, and so on.
3. Blind cavities and narrow slots in dyes can be produced.
4. Preparation of micromolds for injection molding.
5. Microtools having L/D ratio up to 20 can be produced.

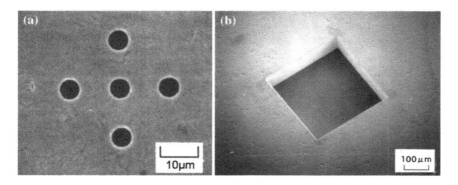

FIGURE 4.19 Microholes produced by EDMM: (a) circular, (b) rectangular. (Copied from [16].) EDMM, electrodischarge micromachining.

FIGURE 4.20 Complex profiles by EDMM: (a, b) rectangular profiles with internal protrusions, (c) cylindrical hole with internal designs. (Copied from [16].) EDMM, electrodischarge micromachining.

4.3.5 Laser Beam Micromachining

Laser beam micromachining (LBMM) is one of the advanced machining methods that utilizes the energy from high-velocity coherent photon beam. The energy released from the photon beam is focused on the target spot, which is transformed into thermal energy to perform machining. The term LASER named as light amplification by stimulated emission of radiation. In 1917, Albert Einstein depicted the relationship between the wavelength of light and atomic energy levels. With the help of this work, the concept of simultaneous and stimulated emission was developed. T.H. Maiman invented the Ruby laser with the support of Einstein's theory in the 1960s at the Hughes Research Laboratories. From this, so many types of lasers were introduced using new working principles and new materials [7].

4.3.5.1 Working Principle

In an atomic model, there are two energy levels such as ground level and higher level (lower and upper) with energy difference $h\nu$. At absolute zero temperature, the electrons in an atom are at ground level and possess the lowest potential energy. By supplying energy from external sources to the electrons at ground level, the electrons will absorb the energy and get excited to higher levels of energy. With this absorption at elevated temperatures, the electron at lower level receives the energy of the photon ($h\nu$) for boosting to an upper level. At this upper-level electron, an unstable energy state is obtained. Hence, it returns to its original state in very short time duration (nanoseconds) by releasing the same photon energy ($h\nu$) gained during absorption. This phenomenon is called spontaneous emission. This emitted photon contains the same frequency as that of an excited photon [19].

Since every atom contains a greater number of electrons, by simultaneous absorption of energy from an external source and release photon energy, the electrons excited to upper level; this is called population inversion. At the upper level, all electrons will obtain an unstable state and return to the ground state by spontaneous emissions. As a result, the emitted photons having the same temporal and special coherence form stimulated emission (as shown in Figure 4.21). This stimulated emission leads to laser formation.

Every laser production crystal consists of two mirrors: one is perfectly reflected and other is partially reflected mirrors. When an excited photon goes toward the totally reflected mirror, it gets reflected once again to stimulate another photon of same temporal and special coherence. Reflected photon coherence passes through partially reflected mirror to generate a straight laser beam. Change of working medium and crystal materials for lasing action produces a different kind of lasers such as solid state and gas lasers (as shown in Figure 4.22) [71].

- Solid-state lasers
  - Ruby laser (Cr–Al Alloy)
  - Nd–glass laser (neodymium)
  - Nd–YAG laser (neodymium-doped with yttrium aluminum garnet)
- Gas lasers
  - Helium–neon laser

Micromachining

- Argon laser
- CO$_2$ laser

Lasers can be divided into two types depending on the power supply nature: continuous-wave lasers and pulsed lasers. When the laser beam emitted from the crystal is continuous, then it is called continuous laser; if the laser beam is emitted in the form of pulses, then it is called pulsed laser. Pulsed lasers had many advantages over continuous-wave lasers in micromachining field since the beam of energy will impact in a short time in the form of laser energy burst. Also, it is often used to machine very hard materials such as ceramics [83,87].

4.3.5.2 Mechanism of Material Removal

In LBMM, electrical energy is supplied as input to the laser generator, which converts electrical energy into high photon energy laser beam. This monochromatic intense narrow laser beam is used for machining of workpiece material. As the beam focuses on the workpiece material, the photon energy gets converted into heat energy to generate local melting and evaporation of focused spot. Hence, LBMM is a thermal machining process where the mechanism of material removal is melting and evaporating (Figure 4.23).

The material removal is basically due to melting and evaporating the workpiece material exposed to severe heat under the high-energy laser beam. This produces

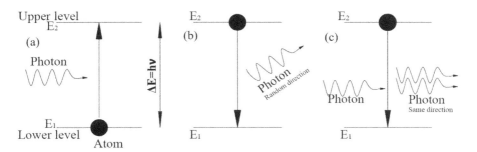

FIGURE 4.21 (a) Absorption, (b) simultaneous emission, (c) stimulated emission.

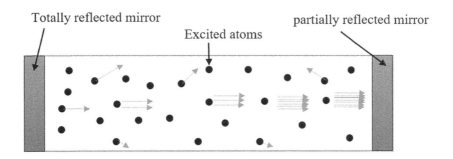

FIGURE 4.22 Lasing action.

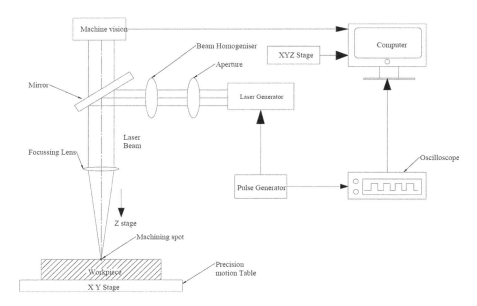

FIGURE 4.23 Schematic diagram of LBMM. (Source: [17].) LBMM, laser beam micromachining.

removal of the material layer by layer under high temperature causing metallurgical changes in workpiece material. Development of recast later, HAZ, and microcracks affect the machined surface. Mechanism of material removal and various layers formed during machining are shown in Figure 4.24 [9,53,54].

By considering these defects, very short pulse durations in the range of pico and femto are developed. Because of the very short duration of the pulse, the

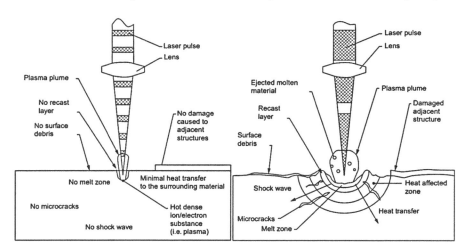

FIGURE 4.24 Laser ablation effect for different pulse durations: femtoseconds and picoseconds (a), nanoseconds and more (b). (Copied from [56].)

Micromachining

laser beam is unable to focus on the same layer of surface for longer duration, resulting in very accurate machining cavity when compared with micro- and nanopulse duration lasers. In recent years, short and ultrashort pulsed lasers such as Cr–LiSCaF (90 fs), Ti–sapphire (6–150 fs), Yb–KGW (176 fs), Yb–YAG (340–730 fs), Nd –YVO$_4$ (2.8–7.9 ps), and microchip lasers (100 ps) were developed for precision machining applications. For rapid cutting and machining applications, a general type of lasers such as Nd–YAG (100–10 ns) [18] or CO$_2$ (50–100 ns) was used.

Particularly for micromachining applications, short and ultrashort lasers are used due to the very short pulse duration [41,88]. Also, to avoid the formation of HAZ and microcracks, excimer and ultrashort lasers were used. The most commonly used laser materials along with wavelength, pulse duration, and frequency are listed in Table 4.2 [90,94].

4.3.5.3 Laser Mask Projection Technique

This technique is suitable only if the beam is larger than the shape of the mask. A stationary mask is used to project a laser beam on the workpiece so that the mask shape is ablated on the workpiece. By using this method, complex 3D shapes can be produced easily on the desired target. The size of the incident beam can be reduced by using a projection lens between mask and target for micromachining applications. A lesser HAZ and greater control of depth can be achieved with this technique (Figure 4.25) [70].

4.3.5.4 Effect of Laser Beam Intensity

Beam intensity is defined as the beam power transferred per unit area perpendicular to the direction of energy emission. Depending on the laser beam intensity, it is used for different applications. If this value is in the order of 10^3 W/cm^2, it is used to warm up the workpiece and laser markings on very soft materials. On further increase in

TABLE 4.2
Commonly Used Lasers for Micromachining Applications

Laser Material	Wavelength (nm)	Pulse Duration	Frequency
Nd–YAG	266–532	100–10 ns	50 Hz
Nd–YVO$_4$	1064	2.8–7.9 ps	84 MHz–77 GHz
Yb–YAG	1030	340–730 fs	35–81 MHz
Yb–KGW	1037	176 fs	86 MHz
Ti–sapphire	750–880	6–150 fs	15 MHz–2 GHz
Cr–LiSCaF	860	90 fs	140 MHz
Fiber lasers	1064	100 ns	20–50 Hz
Microchip lasers	1064	100 ps	100 kHz
ArF (excimer)	193	5–25 ns	1–1000 Hz
Copper vapor lasers	611–578	30 ns	4–20 Hz
KrF (excimer)	248	2–60 ns	1–500 Hz

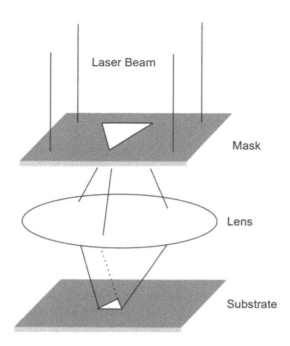

FIGURE 4.25 Mask projection technique for micromachining [70].

the orders of 10^4 and 10^5 W/cm^2, the beam is used for heat treatment applications. With increase in beam intensity further, it is used for melting and cutting operations. Increase in beam intensity will increase the MRR since the energy incident on target from the laser beam will increase. Moreover, all the process parameters and performance measures of LBMM are shown in Figure 4.26 [47,48,69].

4.3.5.5 Material Removal Rate in Pulsed Laser [75]

The material removal mechanism is melting and vaporizing by converting highly accelerated laser beam energy into thermal energy [86]. A model for MRR is developed for drilling a hole by thermal-based LBMM. The diameter of the spot d_s is given by

$$d_s = F_l \times \theta \tag{4.8}$$

where F_l is the focal length of the lens and θ is the beam divergence angle (rad) (Figures 4.27). The area of the laser beam at the focal point, A_s, is

$$A_s = \frac{\pi}{4} \times (F\theta)^2 \tag{4.9}$$

Laser beam power L_p (in Watt) is given by
　where
　E_s = Energy of the laser (in joules)
　Δt = Pulse duration (in s)

Micromachining

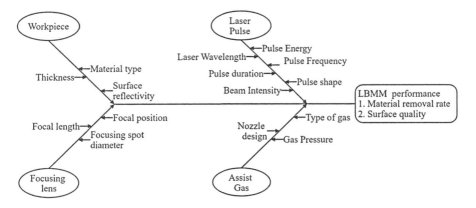

FIGURE 4.26 Cause and effect diagram of LBMM. LBMM, laser beam micromachining.

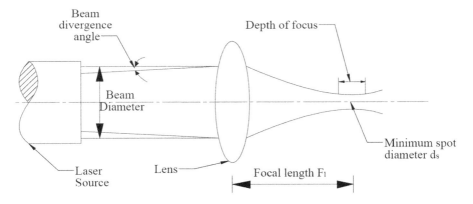

FIGURE 4.27 Focus of laser beam.

$$L_P = \frac{E_S}{\Delta t} \qquad (4.10)$$

Power density or power intensity P_d (in W/mm²) is given by,

$$P_d = \frac{L_p}{A_s} = \frac{4L_p}{\pi(F\theta)^2} \qquad (4.11)$$

LBMM occurs when the power intensity of the laser beam is more than the amount of intensity lost by thermal conduction, convection, and radiation.

The drilling feed rate f (in mm/s) can be described as follows:

$$f = \frac{(C_l L_P)}{(E_V A_s)} = \frac{(C_l P_d)}{E_v} \qquad (4.12)$$

where C_l is a constant (conversion efficiency) depending on the laser material and E_v is vaporization energy of the workpiece material (J/mm³).

The MRR (in mm³/s) can be expressed as [75]

$$\text{MRR} = f \times A_s = \frac{(C_l L_P)}{E_v} \quad (4.13)$$

4.3.5.6 Advantages
1. Microholes can be drilled on difficult-to-cut and hard materials.
2. Complex shapes can be produced easily using CNC assistance.
3. Machining accuracy is high using ultrashort lasers.
4. Ceramic and nonmetallic materials can be machined using pulsed lasers.

4.3.5.7 Limitations
1. HAZ and microcracks will form using continuous-wave gas lasers.
2. Heat-sensitive materials are difficult for laser processing.
3. Efficiency of the process is very less.
4. Initial and maintenance cost is high. Specific power consumption is also high.

4.3.5.8 Applications
In the current era, laser applications are tremendously expanded to many industries such as biomedical, aerospace, automobile, manufacturing, electronics, MEMS, and so on (Figure 4.28) [33,39]. Depending on applications, different types of lasers were used as shown in Table 4.3.

1. This method is used for drilling microholes about 5 μm range.
2. This method is used for producing complex microprofiles.

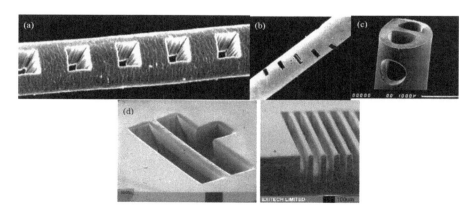

FIGURE 4.28 Applications of LBMM from (a) microchannels (40 μm) drilled on human hair, (b) microholes drilled on PaCO₂ fiber for arterial blood gas sensor by excimer laser (ArF), (c) microholes drilled on PVC bilumen catheter for blood monitoring in new-born babies by excimer laser (ArF) [22]. LBMM, laser beam micromachining.

Micromachining

TABLE 4.3
Applications of Different Types of Lasers

Application	Type of Laser
Trepanned holes	Nd–YAG, CO_2
Holes—diameter <0.25 mm	Ruby, Nd–glass, Nd–YAG
drilling—punching	Nd–YAG, ruby
Thick cutting	CO_2 with gas assistance
Thin slitting of metals	Nd–YAG
Thin slitting of plastics	CO_2
Plastics	CO_2
Metals	Nd–YAG, ruby, Nd–glass
Organics, nonmetal	Pulsed CO_2
Ceramic	Pulsed CO_2, Nd–YAG

3. Laser turning, laser milling, and laser welding can be developed.
4. This method is used for producing microchannels and microgrooves.

4.3.6 ELECTRON BEAM MICROMACHINING

EBMM process is also a thermal process which is like LBMM process. The main difference between these two processes is instead of a laser beam, an electron beam is employed for machining. Additionally, the complete machining environment is done under vacuum chamber, whereas in LBMM, vacuum assistance is not mandatory. The mechanism of material removal in both methods is melting and vaporization. The history and evolution of electron beam micromachining year by year is shown in Table 4.4.

4.3.6.1 Working Principle

Working principle of electron beam machining is that during contact between high velocity-focused electron beam and workpiece, the kinetic energy of the high-velocity electron beam is transformed into heat energy. With the application

TABLE 4.4
History of EBMM [75]

Year	Event
1858	Electron was found from glow
1885	Electron was found from X-ray
1938	EBMM in melting/vaporizing material
1950	Reached adequately high intensity 10^8 W/cm^2 for industrial applications
1958	Steigerwald designed a prototype electron beam equipment that has been built by Messer Griessheim in Germany for welding applications
1959	Used in micromachining

EBMM, electron beam micromachining

of this phenomenon, machining of a target spot can be accomplished because the heat energy released from the beam is greatly higher than the latent heat of the workpiece material; hence, it is high enough to melt and vaporize the spot. The step-by-step mechanism for drilling a hole through EBMM is shown in Figure 4.29. The dotted material present below the workpiece is an auxiliary material for supporting the workpiece and controls the kerf width as well as the shape of the hole.

4.3.6.2 Electron Beam Micromachining Equipment

The schematic diagram of EBMM equipment with all components is shown in Figure 4.30. The high-velocity electrons generated by the electron gun pass through the anode to become high-intense electron beam. This beam is passed through a series of magnetic lenses to converge all the electrons, acquiring a powerful electron beam, which will be focused on the exact spot with the help of deflection coils. This high-power beam transforms from kinetic energy into thermal energy to melt and vaporizes the local surface spot. The workpiece mounted on a CNC-assisted table for producing various geometrical feature designs [14].

4.3.6.3 Components of Electron Beam Micromachining

Electron gun (Figure 4.31): This is the most crucial component of the EBMM which generates very high accelerated free electrons. To generate electrons, very high voltage supply around 150 kV is required. The electron gun consists of a cathode material made by using tungsten or tantalum to withstand very high temperatures. Cathode filaments are heated around 2500°C–3000°C; this high temperature causes thermoionic emission of free electrons from the heated filament. Since cathode cartridge is negatively biased, the thermoionic electrons forcefully drive away from the filament of the cathode. After the cathode, an annular biased grid is connected to the negative bias so that electrons generated from the cathode did not diverge from their path. When the accelerated electrons pass from anode, they will achieve velocity more than 2,00,000 km/s (Figure 4.31).

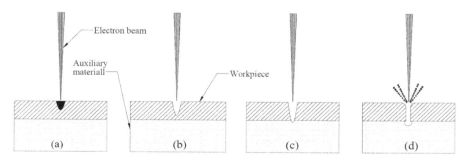

FIGURE 4.29 Material removal mechanism for drilling a hole by EBMM: (a) highly accelerated beam strikes for local heating and target spot surface melting. (b) Increasing the depth of metal vapor capillary action. (c) Higher power density for the beam to penetrate till auxiliary support. (d) Molten metal is expelled due to high vapor pressure. EBMM, electron beam micromachining.

Micromachining

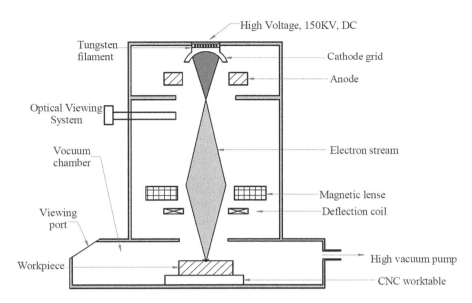

FIGURE 4.30 Schematic diagram of the EBMM process. CNC, computer numerical controller; EBMM, electron beam micromachining.

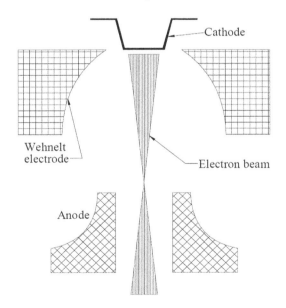

FIGURE 4.31 Structure of thermoelectric gun.

1. **Vacuum chamber**: The complete equipment of EBMM is performed under vacuum (10^{-4}–10^{-5} torr). Vacuum chamber is essential in order to restrict filament oxidation, to prevent the molten metal vapor contamination from air, and to prevent collision of massive air molecules (O_2 and N_2) with

high-velocity electrons, which leads to loss of kinetic energy of electrons. Vacuum can be supplied continuously to the EBMM equipment using a diffusion pump.

2. **Magnetic lens**: Before electron beam reaching the target, magnetic lenses are employed to focus the beam into any desired diameter of 12–25 µm to a precise location on the workpiece. Hence the focused beam gains extremely high-power intensity that will be enough to vaporize any workpiece material. The direction of the beam is governed by Lorentz force law, which is the same principle used in CRT (cathode ray tube).

$$F = q[E + (v \times B)]$$

where F is Lorentz force vector on an electron, E and B are electric and magnetic fields, respectively, and v is velocity vector of the electron.

3. **Deflection coil**: Just below the magnetic lenses, deflection coils are mounted to deflect the beam onto the desired place of the target surface. Any geometrical pattern variations are dependent on the deflection coil system. The beam can be deflected to very small spots in the order of 10–100 µm.

4.3.6.4 Process Parameters (Figure 4.32)

Some of the important process parameters are mentioned in the following; along with these parameters the EBMM performance also depends on thermal properties of the workpiece (such as thermal conductivity, specific heat, melting point, and so on)

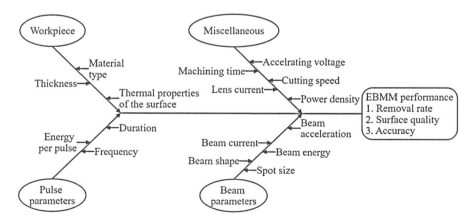

FIGURE 4.32 Cause and effect diagram of EBMM process. EBMM, electron beam micromachining.

Micromachining

1. **Accelerating voltage**: To generate electrons from tungsten or tantalum filament, higher accelerating voltages in the range 150 kV are to be applied between cathode and anode. By increasing the accelerating voltage, beam power will increase, and the electric field inside the electron gun will increase, subsequently causing an increase in beam current and the emission of electrons.
2. **Beam current**: This parameter is mainly related to the number of electrons emitted from the cathode; it can be adjusted in the range of 100 μA to 1 A. By increasing the beam current, the energy per pulse is directly increased.
3. **Pulse duration**: This parameter decides the diameter and depth of the hole. For longer pulse durations, the deeper the hole depth, the wider the hole diameter. This will also cause variation in the recast layer and HAZs to some extent. As already discussed in LBMM, shorter pulse duration gives better results in avoiding thermal effects. Pulse duration can be achieved between as short as 50 μs to as long as 15 ms.
4. **Power density**: This can be defined as emitted beam power per unit area of the spot, and this parameter directly influences MRR. By increasing power density for lower spot size, the size of the hole would be smaller, and MRR will increase.
5. **Lens current**: This parameter helps to find out the working distance (distance between the electron beam gun and the focal point) and also helps to determine the size of the focused spot on the workpiece. By choosing various focal positions, the hole machined can be straight, tapered, bell-shaped, or inversely tapered.

The numerical values of EBMM process parameters for different materials (steel, tungsten, aluminum, alumina, and so on) for drilling and slot cutting are tabulated in Tables 4.5 and 4.6.

4.3.6.5 Advantages
1. Any material can be processed irrespective of electrical or thermal conductivity.
2. Very small size holes (10 μm) can be produced.
3. Surface finish produced is very good and no need of further processing.
4. Complex shapes can be produced by deflecting coils and CNC table assistance.
5. Highly reactive metals such as Al and Mg can be machined easily.

4.3.6.6 Limitations
1. Capital equipment cost is high.
2. Holes of thickness >0.13 mm are tapered and limited to 10 mm thickness.
3. Maintaining a perfect vacuum is difficult.
4. Thermal effects such as recast layer and HAZ will occur on the machined surface.
5. Auxiliary backing material is required.

TABLE 4.5
EBMM Drilling

Material	Working Thickness (mm)	Hole Diameter (μm)	Drilling Time (s)	Accelerating Voltage (kV)	Beam Current (μA)
Tungsten	0.25	25	≤1	140	50
Stainless steel	2.5	125	10	140	100
Stainless steel	1.0	125	≤1	140	100
Aluminum	2.5	125	10	140	100
Alumina (Al_2O_3)	0.75	300	30	125	60
Quartz	3.0	25	≤1	140	10

EBMM, electron beam micromachining.

TABLE 4.6
EBMM Slot Cutting

Material	Working Thickness (mm)	Slot Width (μm)	Cutting Speed (mm/min)	Accelerating Voltage (kV)	Beam Current (μA)
Stainless steel	0.175	100	50	130	50
Tungsten	0.05	25	125	150	30
Brass	0.25	100	50	130	50
Alumina	0.75	100	600	150	200

EBMM, electron beam micromachining.

4.3.6.7 Applications

1. This method is most suitable for drilling of microholes at very high rates.
2. This method is used for producing very small holes in diesel injection nozzles, air brakes, and so on.
3. Electron beam polishing and welding are also important applications (Figure 4.33 and 4.34).

4.3.7 PLASMA ARC MICROMACHINING

Plasma is an ionized gas which will produce high temperatures in the range of 33,000°C. With the help of high-temperature ionized gas, any material can be melted easily. The first plasma state came to existence in the 1900s. But the development of commercial plasma was in 1955 for metal cutting applications.

Micromachining

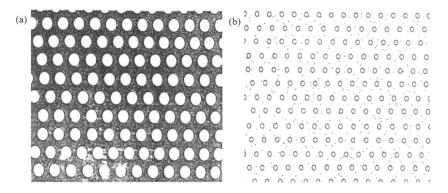

FIGURE 4.33 Pattern of holes drilled by EBMM. (a) 90-μm-diameter holes drilled on 0.2-mm-thickness stainless steel sheet with density 4000 holes/cm², time/hole:10 μs. (b) 6-μm-diameter holes drilled on 12-μm-thickness synthetic fabric sheet with density 20,000 holes/cm², time/hole: 2 μs. (Steigerwald and Mayer (1967).) EBMM, electron beam micromachining.

FIGURE 4.34 Microholes in diesel injection nozzle [81].

4.3.7.1 Working Principle

Plasma is an ionized gas generated at high temperatures used to melt the surface of the workpiece so that the material gets evaporated; hence, the mechanism of material removal is melting and evaporating. Plasma arc micromachining (PAMM) as shown in Figure 4.35 consists of a plasma torch with an electrode holder and water-cooled nozzle. Power supply is given between workpiece and plasma generator so that the plasma arc is generated in the IEG. This method is well suited for sheet-cutting operations, heat treatment, plasma arc welding, and so on.

4.3.7.2 Process Parameter

The quality of the material surface and cutting rate depends on various factors in PAMM. Some of the factors that will influence the process are as follows:

- Nozzle diameter
- Gas mixture composition
- Electrode standoff distance
- Plasma jet feed rate
- Plasma gas pressure

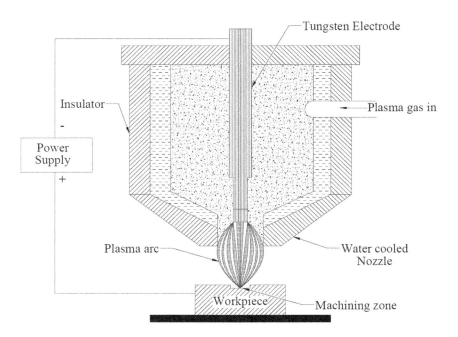

FIGURE 4.35 Schematic diagram of PAMM. PAMM, plasma arc micromachining.

4.3.7.3 Advantages
1. MRR is faster.
2. Hard and difficult-to-cut materials are processed easily.
3. Process variables are controlled easily.

4.3.7.4 Limitations
1. Initial cost is very high.
2. Safety precautions are more.
3. Machined surface contains metallurgical changes due to excessive heat.

4.3.7.5 Applications
1. The main application of this process is profile cutting on difficult-to-cut materials.
2. This method is used for plasma arc welding for joining dissimilar metals.
3. This method is used for plasma arc surfacing for any heat treatment process and plasma arc spraying for molten metal coatings.
4. Turning and milling can be performed by using plasma.

4.4 CONCLUSIONS AND FUTURE STUDY

The micromachining processes with physical aspects of working mechanism and conceptual understanding are stated in this chapter. Moreover, a comparison table is drawn by considering various factors of all micromachining processes (as shown in

Micromachining

Table 4.7). The literature gaps and conclusions drawn from this work for future study are listed in the following:

1. This chapter addressed the important aspects of material removal mechanisms, history of the process, energy sources, working principles, effects of process parameters, components in machining system, advantages, limitations, and applications of each micromachining process individually.
2. Conventional micromachining requires great strengths (greater than ultimate strength of the workpiece material) to initiate shear on the processing surface to initiate this large amount of rotational speeds (minimum 60,000 rpm) required. Though these processes have more MRR, the vibration chatter, poor dimensional accuracy, tool wear, and so on limited the application of these processes.
3. The variation of performance measures such as surface quality and MRR with different input variables was studied. Electrical process parameters and pulse-related parameters had more influence when compared with other in EDMM and ECMM. The variability in machining process for type of application also reported. Some of these are laser mask projection technique, laser milling, laser drilling, laser cutting, plasma cutting, wire-EDM, reverse EDM, EDG, powder-mixed EDM, wire-ECM, AWJMM, and so on.
4. Environmental aspects of the micromachining process have not been addressed a lot, and there is a scope for future study. Also the micro/nanofinishing methods are in trend to carry forward the analysis of micromachined surface under scanning electron microscope (SEM) and transmission electron microscope (TEM) for microstructural observation.
5. There is a massive scope in exploring the combination of the different processes (based on different energy sources such as mechanical, thermal, electrothermal, chemical, electrochemical) to develop hybrid micromachining processes to achieve desired accuracy and removal rate. Deep studies are needed to understand the mechanisms behind the hybrid processes; to investigate the machining performance of these processes, simulation with various parameter ranges can be done.
6. The surface quality (regarding HAZs, microcracks, resolidified layers, and residual stressed surface) of the machined surface particularly with thermoelectrical energy–related micromachining processes such as micro-EDM, micro- LBM, micro-EBM, and micro-PAM need to be improved with further research.
7. Most of the users face difficulty in holding the micro- and tiny parts. So there is a need of development in work handling system that has to be upgraded to the robotic gripping for accurate holding or better work holding and positioning system. Preparation of miniature tools, microtool holders, and handling system with precision is still a challenge in micromachining field.
8. A few limitations of the machining process can be reduced by simulating the process variables with current computer-based modeling techniques. Even though previous literatures are reported to some extent, modeling and simulation coupled with experimental investigations give better results and still need to be explored for every micromachining process.

TABLE 4.7

Comparison of All Micromachining Processes

Process	Energy Type	Mechanism of MRR	Medium	Energy Source	Tool Wear	Workpiece Material	HAZ	Machine Cost	Surface Finish	Efficiency
USMM	Mechanical	Brittle fracture	Abrasive slurry	Abrasive impact loads	High	Hard and brittle materials	No	Low	Good	High
AJMM	Mechanical	Erosion or abrasion	Abrasive+air mixture	Pneumatic pressure	No	Metals, ceramics, plastics, etc.	No	Medium	Poor	High
ECMM	Electrochemical	Ion displacement	Conductive electrolyte	High current	No	Electrically conductive materials	No	Medium	Very good	Low
EDMM	Thermoelectric	Spark erosion	Dielectric fluid	High voltage	Low	Melting and vaporization	Yes	Low	Good	High
LBMM	Thermoelectric	Melting and vaporization	Radiation	High-velocity laser beam	No	Any material melts on heating	Yes	Very high	Very good	Very high
EBMM	Thermoelectric	Melting and vaporization	Vacuum	High accelerated electron beam	No	Any material melts on heating	Yes	Very high	Very good	Very high
PAMM	Thermoelectric	Melting and vaporization	Argon-hydrogen, air or nitrogen	Ionized material	No	Electrically conductive	Yes	Very high	Good	Low
Conventional MM	Mechanical	Shear	Coolant	Micro cutting tool	Very high	Any material except hard materials	No	Low	Poor	High

AJMM, abrasive jet micromachining; EBMM, electron beam micromachining; ECMM, electrochemical micromachining; EDMM, electrodischarge micromachining; HAZ, heat-affected zone; LBMM, laser beam micromachining; PAMM, plasma arc micromachining; USMM, ultrasonic micromachining.

REFERENCES

1. K. Abhishek, S. S. Hiremath, and S. Karunanidhi. A novel approach to produce holes with high degree of cylindricity through Micro-Abrasive Jet Machining (μ-AJM). *CIRP Journal of Manufacturing Science and Technology*, 21:110–119, 2018.
2. F. Ahmadzadeh, S. S. H. Tsai, and M. Papini. Effect of curing parameters and configuration on the efficacy of ultraviolet light curing self-adhesive masks used for abrasive jet micro-machining. *Precision Engineering*, 49:354–364, 2017.
3. Y. Ahn and S. H. Lee. Classification and prediction of burr formation in micro drilling of ductile metals. *International Journal of Production Research*, 7543(June):1–14, 2017.
4. A. A. A. Aliyu, A. M. Abdul-Rani, T. L. Ginta, C. Prakash, E. Axinte, M. A. Razak, and S. Ali. A review of additive mixed-electric discharge machining: current status and future perspectives for surface modification of biomedical implants. *Advances in Materials Science and Engineering*, 01:1–23, 2017.
5. P. S. Bains, S. S. Sidhu, and H. S. Payal. Fabrication and machining of metal matrix composites: a review. *Materials and Manufacturing Processes*, 31(5):553–573, 2016.
6. B. Bhattacharyya, J. Munda, and M. Malapati. Advancement in electrochemical micromachining. *International Journal of Machine Tools and Manufacture*, 44(15):1577–1589, 2004.
7. A. Bucciarelli, P. D. Kuila, S. N. Melkote, and A. Fortunato. Micro-machinability of A-286 steel with and without laser assist. *Procedia CIRP*, 46:432–435, 2016.
8. J. B. Byiringiro, T. J. Ko, H. C. Kim, and I. H. Lee. Optimal conditions of SU-8 mask for micro-abrasive jet machining of 3-D freeform brittle materials. *International Journal of Precision Engineering and Manufacturing*, 14(11):1989–1996, 2013.
9. G. B. J. Cadot, D. A. Axinte, and J. Billingham. Continuous trench, pulsed laser ablation for micro-machining applications. *International Journal of Machine Tools and Manufacture*, 107:8–20, 2016.
10. H. C. Kim, I. H. Lee, and T. J. Ko. 3D tool path generation for micro- abrasive jet machining on 3D curved surface. *International Journal of Precision Engineering and Manufacturing*, 14(9):1519–1525, 2013.
11. F. Chen, X. Miao, Y. Tang, and S. Yin. A review on recent advances in machining methods based on abrasive jet polishing (AJP). *International Journal of Advanced Manufacturing Technology*, 90(1–4):785–799, 2017.
12. X. L. Chen, B. Y. Dong, C. Y. Zhang, H. P. Luo, J. W. Liu, Y. J. Zhang, and Z. N. Guo. Electrochemical direct-writing machining of micro-channel array. *Journal of Materials Processing Technology*, 265:138–149, 2019.
13. S. F. Chuang, Y.-C. Chen, and W.-T. Chang. Nondestructive testing and evaluation nondestructive web thickness measurement of micro-drills with an integrated laser inspection system. *Nondestructive Testing and Evaluation*, (January):37–41, 2015.
14. H. Cui, R. Wang, D. Wei, J. Huang, J. Guo, and X. Li. Surface modification of the carbon tool steel by continuous scanning electron beam process. *Nuclear Instruments and Methods in Physics Research, B*, 440(November 2018):156–162, 2019.
15. S. Das, B. Doloi, and B. Bhattacharyya, "Recent Advancement on Ultrasonic Micro Machining (USMM) Process," in Non-traditional Micromachining Processes, 61–91, 2017.
16. J. Paulo Davim, Machining of Complex Sculptured Surfaces. *Springer-Verlag London*, 53(9), 2015.
17. S. Debnath, S. Kunar, S. S. Anasane, and B. Bhattacharyya, "Non-traditional Micromachining Processes : Opportunities and Challenges," in Materials Forming, Machining and Tribology, 1–59, 2017.
18. A. K. Dubey and V. Yadava. Experimental study of Nd: YAG laser beam machining — an overview. *Journal of materials processing technology*, 5:15–26, 2007.

19. A. K. Dubey and V. Yadava. Laser beam machining — a review. *International Journal of Machine Tools & Manufacture*, 48:609–628, 2008.
20. P. Ervine, G. E. Odonnell, and B. Walsh. Fundamental investigations into burr formation and damage mechanisms in the micro-milling of a biomedical grade polymer. *Machining Science and Technology*, 19(1):112–133, 2015.
21. S. Feng, C. Huang, J. Wang, H. Zhu, P. Yao, and Z. Liu. An analytical model for the prediction of temperature distribution and evolution in hybrid laser-waterjet micromachining. *Precision Engineering*, 47:33–45, 2017.
22. M. C. Gower. Industrial applications of pulsed laser micromachining. *Conference on Lasers and Electro-Optics Europe – Technical Digest*, 7(2):247, 1998.
23. B. Guimarães, D. Figueiredo, C. M. Fernandes, F. S. Silva, G. Miranda, and O. Carvalho. Laser machining of WC-Co green compacts for cutting tools manufacturing. *International Journal of Refractory Metals and Hard Materials*, 81(February):316–324, 2019.
24. N. Haghbin, F. Ahmadzadeh, and M. Papini. Masked micro-channel machining in aluminum alloy and borosilicate glass using abrasive water jet micro- machining. *Journal of Manufacturing Processes*, 35(October 2017):307–316, 2018.
25. N. Haghbin, F. Ahmadzadeh, J. K. Spelt, and M. Papini. High pressure abrasive slurry jet micro-machining using slurry entrainment. *International Journal of Advanced Manufacturing Technology*, 84(5–8):1031–1043, 2016.
26. R. H. M. Jafar, H. Nouraei, M. Emamifar, M. Papini, and J. K. Spelt. Erosion modeling in abrasive slurry jet micro-machining of brittle materials. *Journal of Manufacturing Processes*, 17:127–140, 2015.
27. M. Hasan, J. Zhao, and Z. Jiang. A review of modern advancements in micro drilling techniques. *Journal of Manufacturing Processes*, 29:343–375, 2017.
28. R. H. M. Jafar, M. Papini, and J. K. Spelt. Simulation of erosive smoothing in the abrasive jet micro-machining of glass. *Journal of Materials Processing Technology*, 213(12):2254–2261, 2013.
29. S. Jahanmir, Z. Ren, H. Heshmat, and M. Tomaszewski. Design and evaluation of an ultrahigh speed micro-machining spindle. *Machining Science and Technology*, 14(2):224–243, 2010.
30. S. James. Experimental study on micromachining of CFRP/Ti stacks using micro ultrasonic machining process. *International Journal of Advanced Manufacturing Technology*, 95:1539–1547, 2018.
31. J. Kumar. Investigations into the surface quality and micro-hardness in the ultrasonic machining of titanium (ASTM GRADE-1) surface roughness. *The Brazilian Society of Mechanical Sciences and Engineering*, 36:807–823, 2014.
32. S. K. Jui, A. B. Kamaraj, and M. M. Sundaram. High aspect ratio micromachining of glass by electrochemical discharge machining (ECDM). *Journal of Manufacturing Processes*, 14:460–466, 2013.
33. A. Kaldos, H. J. Pieper, E. Wolf, and M. Krause. Laser machining in die making – A modern rapid tooling process. *Journal of Materials Processing Technology*, 155–156(1–3):1815–1820, 2004.
34. Kibria, G., Bhattacharyya, B., Paulo, J., & Editors, D. (2017). Non-traditional Micromachining ProcessesMaterials Forming, Machining and Tribology, 1–422.
35. E. Krebs, M. Wolf, D. Biermann, W. Tillmann, and D. Stangier. High-quality cutting edge preparation of micromilling tools using wet abrasive jet machining process. *Production Engineering*, 12(1):45–51, 2018.
36. A. Kumar and S. S. Hiremath. Improvement of geometrical accuracy of micro holes machined through micro abrasive jet machining. *Procedia CIRP*, 46:47–50, 2016.

37. A. Kumar and D. Partha. Machining of circular micro holes by electrochemical micro-machining process. *Advances in Manufacturing Science and Technology*, 1:314–319, 2013.

38. S. Kumar, A. Dvivedi, and P. Kumar. On tool wear in rotary tool micro-ultrasonic machining tool rotation tool wear. *Proceedings of the 3rd Pan American Materials Congress*, 2017.

39. S. M. Langan, D. Ravindra, and A. B. Mann. Mitigation of damage during surface finishing of sapphire using laser-assisted machining. *Precision Engineering*, 56:1–7, 2019.

40. S. P. Lee, H. W. Kang, S. J. Lee, I. H. Lee, T. J. Ko, and D. W. Cho. Development of rapid mask fabrication technology for micro-abrasive jet machining. *Journal of Mechanical Science and Technology*, 22(11):2190–2196, 2008.

41. J. Lehr and A. M. Kietzig. Production of homogenous micro-structures by femtosecond laser micro-machining. *Optics and Lasers in Engineering*, 57:121–129, 2014.

42. Z. Li, J. Bai, Y. Cao, Y. Wang, and G. Zhu. Fabrication of microelectrode with large aspect ratio and precision machining of micro-hole array by micro-EDM. *Journal of Materials Processing Technology*, 268(June 2018):70–79, 2019.

43. H. S. Lian, Z. N. Guo, J. W. Liu, Z. G. Huang, and J. F. He. Experimental study of electrophoretically assisted micro-ultrasonic machining. *International Journal of Advanced Manufacturing Technology*, 85:2115–2124, 2016.

44. X. Liu, R. E. DeVor, S. G. Kapoor, and K. F. Ehmann. The mechanics of machining at the microscale: assessment of the current state of the science. *Journal of Manufacturing Science and Engineering*, 126(4):666, 2005.

45. K. Aslantas, M. Percin, A. Cicek, I. Ucun, and Y. Kaynak. Micro-drilling of Ti–6Al–4V alloy: the effects of cooling/lubricating. *Precision Engineering*, 45:450–462, 2016.

46. A. Malik and A. Manna. An experimental investigation on developed WECSM during micro slicing of e-glass fibre epoxy composite. *The International Journal of Advanced Manufacturing Technology*, 85:2097–2106, 2016.

47. K. K. Mandal, A. S. Kuar, and S. Mitra. Experimental investigation on laser micro-machining of Al 7075 Al. *Optics and Laser Technology*, 107:260–267, 2018.

48. M. V. Pantawane, S. S. Joshi, and N. B. Dahotre. Laser beam machining of aluminum and aluminum alloys. *Aluminum Science and Technology*, 2:519–541, 2018.

49. R. Melentiev and F. Fang. Recent advances and challenges of abrasive jet machining. *CIRP Journal of Manufacturing Science and Technology*, 22:1–20, 2018.

50. J. Nam and S. W. Lee. Machinability of titanium alloy (Ti-6Al-4V) in environmentally-friendly micro-drilling process with nanofluid minimum quantity lubrication using nanodiamond particles. *International Journal of Precision Engineering and Manufacturing-Green Technology*, 5(1):29–35, 2018.

51. Y. Zhu, A. Farhadi, L. Gu, X. Kang, and W. Zhao. Observation analysis of arc plasma channel developing and expansion behavior in single arc discharging. *Procedia Manufacturing*, 26:454–461, 2018.

52. A. Nouhi, M. R. Sookhak Lari, J. K. Spelt, and M. Papini. Implementation of a shadow mask for direct writing in abrasive jet micro-machining. *Journal of Materials Processing Technology*, 223:232–239, 2015.

53. Y. Okamoto, A. Okada, A. Kajitani, and T. Shinonaga. High surface quality micro machining of monocrystalline diamond by picosecond pulsed laser. *CIRP Annals – Manufacturing Technology*, 68(1):197–200, 2019.

54. P. Parandoush and A. Hossain. A review of modeling and simulation of laser beam machining. *International Journal of Machine Tools and Manufacture*, 85:135–145, 2014.

55. V. R. Perla, S. K. Suraparaju, K. J. Rathanraj, and A. S. Reddy. Effect of Electrochemical Micromachining Process Parameters on Surface Roughness and Dimensional Deviation of Ti6Al4V by Tungsten Electrode. In Recent Advances in Mechanical Engineering, 211–226. Singapore: Springer, 2020.

56. D. T. Pham, S. S. Dimov, P. V. Petkov, and S. P. Petkov. Laser milling. *Proceedings of the Institution of Mechanical Engineers, Part B: Journal of Engineering Manufacture*, 216(5):657–667, 2002.

57. C. Prakash, H. K. Kansal, B. S. Pabla, and S. Puri. Processing and characterization of novel biomimetic nanoporous bioceramic surface on β-Ti implant by powder mixed electric discharge machining. *Journal of Materials Engineering and Performance*, 24(9):3622–3633, 2015.

58. C. Prakash, H. K. Kansal, B. S. Pabla, and S. Puri. Multi-objective optimization of powder mixed electric discharge machining parameters for fabrication of biocompatible layer on β-Ti alloy using NSGA-II coupled with Taguchi based response surface methodology. *Journal of Mechanical Science and Technology*, 30(9):4195–4204, 2016.

59. C. Prakash, H. K. Kansal, B. S. Pabla, and S. Puri. Powder mixed electric discharge machining: an innovative surface modification technique to enhance fatigue performance and bioactivity of β-Ti implant for orthopedics application. *Journal of Computing and Information Science in Engineering*, 16(4)1–9, 2016.

60. C. Prakash, H. K. Kansal, B. S. Pabla, and S. Puri. Experimental investigations in powder mixed electric discharge machining of Ti–35Nb–7Ta–5Zrβ- titanium alloy. *Materials and Manufacturing Processes*, 32(3):274–285, 2017.

61. C. Prakash, H. K. Kansal, B. S. Pabla, S. Puri, and A. Aggarwal. Electric discharge machining – a potential choice for surface modification of metallic implants for orthopedic applications: a review. *Proceedings of the Institution of Mechanical Engineers, Part B: Journal of Engineering Manufacture*, 230(2):331–353, 2016.

62. C. Prakash, S. Singh, B. S. Pabla, and M. S. Uddin. Synthesis, characterization, corrosion and bioactivity investigation of nano-HA coating deposited on biodegradable Mg-Zn-Mn alloy. *Surface and Coatings Technology*, 346:9–18, 2018.

63. C. Prakash, S. Singh, C. I. Pruncu, V. Mishra, G. Królczyk, D. Y. Pimenov, and A. Pramanik. Surface modification of Ti-6Al-4V alloy by electrical discharge coating process using partially sintered Ti-Nb electrode. *Materials*, 12(7):1006, 2019.

64. C. Prakash, S. Singh, M. Singh, K. Verma, B. Chaudhary, and S. Singh. Multi-objective particle swarm optimization of EDM parameters to deposit HA-coating on biodegradable Mg-alloy. *Vacuum*, 158:180–190, 2018.

65. A. Pramanik, A. K. Basak, and C. Prakash. Understanding the wire electrical discharge machining of Ti6Al4V alloy. *Heliyon*, 5(4):e01473, 2019.

66. A. Pramanik, M. N. Islam, A. K. Basak, Y. Dong, G. Littlefair, and C. Prakash. Optimizing dimensional accuracy of titanium alloy features produced by wire electrical discharge machining. *Materials and Manufacturing Processes*, 34(10):1083–1090, 2019.

67. K. P. Rajurkar, M. M. Sundaram, and A. P. Malshe. Review of electrochemical and electrodischarge machining. Procedia CIRP, 6:13–26, 2013.

68. P. S. V. R. Rao, A. L. Naidu, and S. Kona. Design and fabrication of abrasive jet machine (AJM). *Mechanics and Mechanical Engineering*, 22(4):1471–1482, 2018.

69. V. C. Reddy, G. H. Gowd, and M. L.S. D. Kumar. Empirical modeling & optimization of laser micro – machining process parameters using genetic algorithm. *Materials Today: Proceedings*, 5(2):8095–8103, 2018.

70. N. H. Rizvi. Production of novel 3D microstructures using excimer laser mask projection techniques. *Symposium on Design, Test, and Microfabrication of MEMS and MOEMS*, 3680(April):546–552, 1999.

Micromachining

71. A. N. Samant and N. B. Dahotre. Laser machining of structural ceramics – a review. *Journal of the European Ceramic Society*, 29:969–993, 2009.
72. Saragih A.S., Lee J.H., Ko T.J., Kim H.S. (2007) Thick SU-8 Mask for Micro Channeling of Glass by using Micro Abrasive Jet Machining. In: Towards Synthesis of Micro-/Nano-systems, 187–190. London: Springer, 2006.
73. I. Saxena, S. Wolff, and J. Cao. Unidirectional magnetic field assisted laser induced plasma micro-machining. *Manufacturing Letters*, 3:1–4, 2015.
74. D. Sreehari and A. Kumar. On form accuracy and surface roughness in micro-ultrasonic machining of silicon microchannels. *Precision Engineering*, 53(March):300–309, 2018.
75. S.Y. Liang and A.J. Shih. Analysis of Machining and Machine Tools, 1–277. Boston: Springer, 2016.
76. X. Su, L. Shi, W. Huang, and X. Wang. A multi-phase micro-abrasive jet machining technique for the surface texturing of mechanical seals. *International Journal of Advanced Manufacturing Technology*, 86(5–8):2047–2054, 2016.
77. A. Sun, Y. Chang, and H. Liu. Metal micro-hole formation without recast layer by laser machining and electrochemical machining. *Optik*, 171(June):694–705, 2018.
78. A. Sun, Y. Chang, and H. Liu. Numerical simulation of laser drilling and electrochemical machining of metal micro-hole. *Optik*, 181:92–98, 2019.
79. R. Suresh, K. S. Reddy, and K. Shapur. Abrasive Jet Machining for Micro-hole Drilling on Glass and GFRP Composites. *Materials Today: Proceedings*, 5(2):5757–5761, 2018.
80. I. Tansel, O. Rodriguez, M. Trujillo, E. Paz, and W. Li. Micro-end-milling – I. Wear and breakage. *International Journal of Machine Tools and Manufacture*, 38(12):1419–1436, 1998.
81. Faculdade D E Tecnologia. Caracterização básica de spray demistura ternária diesel-etanol-óleovegetal. Technical report, 2016.
82. M. Tisza, Z. Lukacs, and G. Gal. Self-pierce riveting of three high strength steel and aluminium alloy sheets. *International Journal of Material Forming*, 1:185–188, 2008.
83. T. Uenohara, Y. Takaya, and Y. Mizutani. Laser micro machining beyond the diffraction limit using a photonic nanojet. *CIRP Annals – Manufacturing Technology*, 66(1):491–494, 2017.
84. Y. Uno, A. Okada, K. Uemura, P. Raharjo, S. Sano, Z. Yu, and S. Mishima. A new polishing method of metal mold with large-area electron beam irradiation. *Journal of Materials Processing Technology*, 188:77–80, 2007.
85. V. V. Vanmore and U. A. Dabade. Development of laval nozzle for micro abrasive jet machining MAJM. processes. *Procedia Manufacturing*, 20:181–186, 2018.
86. M. Wu, B. Guo, Q. Zhao, R. Fan, Z. Dong, and X. Yu. The influence of the focus position on laser machining and laser micro-structuring monocrystalline diamond surface. *Optics and Lasers in Engineering*, 105(2):60–67, 2018.
87. X. Wu, L. Li, N. He, G. Zhao, and J. Shen. Laser induced oxidation of cemented carbide during micro milling. *Ceramics International*, 45(12):15156–15163, 2019.
88. Y. Xing, L. Liu, X. Hao, Z. Wu, P. Huang, and X. Wang. Micro-channels machining on polycrystalline diamond by nanosecond laser. *Optics and Laser Technology*, 108:333–345, 2018.
89. H. Zarepour, S. H. Yeo, P. C. Tan, and E. Aligiri. A new approach for force measurement and workpiece clamping in micro-ultrasonic machining. *International Journal of Advanced Manufacturing Technology*, 53(5–8):517–522, 2011.
90. A. Žemaitis, M. Gaidys, P. Gečys, G. Račiukaitis, and M. Gedvilas. Rapid high-quality 3D micro-machining by optimised efficient ultrashort laser ablation. *Optics and Lasers in Engineering*, 114(October 2018):83–89, 2019.

91. S. Zhang, Y. Zhou, H. Zhang, Z. Xiong, and S. To. Advances in ultra-precision machining of micro-structured functional surfaces and their typical applications. *International Journal of Machine Tools and Manufacture*, 142(April):16–41, 2019.
92. S. Zhang, Z. Liang, X. Wang, T. Zhou, L. Jiao, and P. Yan. Grinding process of helical micro-drill using a six-axis CNC grinding machine and its fundamental drilling performance. *International Journal of Advanced Manufacturing Technology*, 86:2823–2835, 2016.
93. L. J. Zheng, C. Y. Wang, and Y. P. Qu. Interaction of cemented carbide micro-drills and printed circuit boards during micro-drilling. *International Journal of Advanced Manufacturing Technology*, 77:1305–1314, 2015.
94. H. Zhu, Z. Zhang, J. Xu, K. Xu, and Y. Ren. An experimental study of micro-machining of hydroxyapatite using an ultrashort picosecond laser. *Precision Engineering*, 54:154–162, 2018.

5 A Review Study on Miniaturization
A Boon or Curse

Ankit Sharma
Chitkara College of Applied Engineering,
Chitkara University, Punjab, India

Vivek Jain, Dheeraj Gupta, and Atul Babbar
Thapar University
Shree Guru Gobind Singh Tricentenary
University, Gurugram, Haryana, India

CONTENTS

5.1 Introduction 111
5.2 Recent Studies 113
    5.2.1 Process Physics 113
    5.2.2 Minimum Chip Thickness and Specific Cutting Energy 115
    5.2.3 Ductile Mode Machining 116
    5.2.4 Edges and Surface Finish 117
    5.2.5 Workpiece and Design Issues 119
    5.2.6 Machines, Tools, and Systems for Micromachining 119
    5.2.7 Cutting Fluid 120
    5.2.8 Machine Components and Controls 122
    5.2.9 Metrology in Micromachining 123
5.3 Conclusion and Brief Discussion 125
5.4 Future Scope 125
References 126

5.1 INTRODUCTION

As time passes, the manufacturing of smaller or microcomponents/parts has become the main center of attraction for the researchers and engineers. According to Dornfeld [1], in the early century, the microproducts have demand in the field of art followed by science and engineering because of their new applications, low cost, high-quality product, and better performance. Conventional as well as non-conventional machining processes have always played a vital role to convert raw material into the finish product. Similarly, the ability of machining at the microlevel

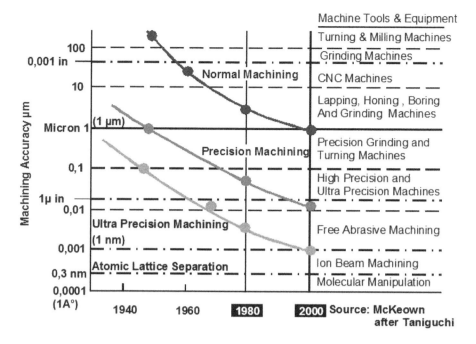

FIGURE 5.1 Micromachining capability versus time [1].

is continuously improved day by day. In the research work of Taniguchi [2], present condition and potential development of ultraprecision machining have been discussed. The processing of micromachining and its related parameters were studied for the past 60 years. In Figure 5.1, micromachining accuracy range with respect to 1940–2000 period was analyzed [1]. Figure 5.1 represents the material machining accuracy during one cycle of the manufacturing process. According to Ghosh et al. [3], as the technology is improved, scientists and researchers are facing a huge number of troubles in the field of fabrication. The brainstorming points for the research developers are low machinability with new materials, precise and dimensional accuracy, and less production along with high productivity.

According to Rajput [4], micromachining is a concern with miniaturization of shape and size of the product by conventional as well as nonconventional machining techniques. In that case, metal removal rate is in microdimensional level. Major applications of micromachining are integrated circuits (ICs), small surgical tools, fuel injection nozzles, microeffectors to handle biological cells, and compact electrical circuits. Vibration is applied to the cutting tools to perform machining without any external force in conventional machining. For abrasive machining, diamond and hard ceramics material tools are generally used. In optical and electronic sectors, microdiamond tool is used with an accuracy of 0.01 μm along with high surface finishing. During ductile mode of machining, diamond fabricates superfinish surface on hard as well as brittle materials. Electronic and magnetic-based materials such as silicon, ceramics, quartz wafers, and so on are fabricated by abrasive microlapping.

Miniaturization—A Boon or Curse

On the other hand, micromachining is also done in large number by nonconventional machining such as micro-EDM (electrical discharge machining), microlaser machining, chemical machining, and thermal erosion.

According to Kalpakjian et al. [5], in the electronics field, materials such as semiconductors were invented in the 1940s. The invention of the transistor set, cellular telephones, and automotive control devices becomes one of the greatest developments in the history of technology. The transistor was the motivated source of development for the complex ICs. Nowadays, ICs have merits such as a high degree of complexity, reduced size, and less cost factor. As the modern production technology came into the picture, the size of the components diminished. Fabrication of a tiny range of microelectronics products takes place in a dust-proof environment. Dirt-free laboratories are broadly used for microfabrication generally for the maximum of 0.5-μm particles per cubic foot. To develop the microparts, MEMS (microelectro-mechanical system) devices are used. MEMSs are consisting of electrical as well as mechanical systems by means of characteristic size, i.e., 1 mm. For the fabrication of electronic components, the batch process technologies are applied even though the alternative process also considered. Some of the extended applications of MEMS devices are precise rapid sensors, artificial organs, ink jet printers, and accelerometers. Alting et al. [6] studied about the microengineering field of microsize product design parameters as well as the various fabrication methods for manufacturing small-scale components.

5.2   RECENT STUDIES

Some of the studies have been done to observe the different factors and parameters of the micromachining techniques. After World War II, a new evolution came, and that was the industrialization. As time moves on, the development in the field of manufacturing inclined drastically. The conventional machining and nonconventional machining processes were generally applied for fabrication. The demand of the market was focused on the machining and manufacturing of hard as well as brittle materials. On the other hand, the demand of the precise, high surface finished, and microdimension products were also increased, especially in the industries such as electronics, biomedical, aircraft, defense, and automobile. Definitely, the miniaturization of the products leads to a large extent of applications. In spite of this, miniaturization is also creating some negative effects on the cost of manufacturing, metrological instrumentation, working stability, and so on. Here the study elaborates that the miniaturization is a boon or curse.

5.2.1   PROCESS PHYSICS

In the research work of Hackert et al. [7], jet electrochemical machining process plays a vital role to generate complex as well as critical shape components by control over the movement of the nozzle and electric current rate. In this machining process, direct current was supplied by an electrolyte jet between the workpiece as an anode and the tool as a cathode. The electrolytic liquid jet was driven at an average velocity of 20 m/s through a small size nozzle. This process has an ambient air medium. The pulses-free pump provides the constant and stable pressure supply of steady flow electrolyte, and it fascinated the well-defined geometrical shape jet. The jet perpendicularly strikes over the workpiece surface. Here, as the current rate increases to some extent, it leads to incline the surface finishing microrange, i.e., 1000 Amp/cm^2. On the other side, geometry and design consideration of the nozzle also helps to get precise machining. As the diameter size of the nozzle reduced up to 50 m, it leads to a decrease in the material removal rate (MRR). By providing flat electrolyte jet, it gives the applications such as grooves and slots while without any nozzle movement [8]. The extended jet shape fascinated the high metal removal rate as compared with the cylindrical shape jet. Multiple electrolyte nozzles provide further improvement in the machining process and also help in increasing its applications.

Jian et al. [9] studied the development related to the ceramic reinforcement matrix material under micromachining. Many of the sectors such as defense, energy resources, medical, automobile, aeronautical, and biotechnology used the material such as metal–matrix composite (MMC) materials. The eye-catching properties of MMC materials are mechanical properties that included light in weight, creep resistance property, low thermal expansion rate, high strength, less fatigue failure, anticorrosion and antioxidation, and good wear resistance. Some of the specific applications of MMC materials while considering their major properties are microsensors, microholes in optical, microfluidic channels for fuel cells, and a micronozzle array for multiplexed electrospray systems and actuators. The MMC materials such as aluminum-based MMCs (Al-MMCs) or magnesium-based MMCs (Mg-MMCs) possess light weight and high toughness property, which gave the good option to manufacture the parts for such applications [10–12]. Under suitable conditions, miniature parts may lead to positive effects on energy efficiency. According to researchers while working on the micromachining process, the small size workpiece can be manufactured more efficiently, which is considered for less economy with high quality. On the other hand, macroscale-level manufacturing is required to concentrate on modeling the process states, parameters related to cutting forces and tool vibration, and also the workpiece dimensional accuracy and surface reliability.

Micromachining techniques have sustained many characteristics related to the conventional machining [1]. As the size of the material goes toward miniaturization, the characteristic of the material remains same up to some extent. Once the ratio between the workpiece size and the tool dimension (width or length) became diminished, the miniaturization of the size may lead to affect the all the parameters such as the depth of cut with respect to the tool edge radius which were related during micromachining. It was also noted that the microstructures of workpiece material were affected by the cutting mechanism. The core difference between the micromachining and micromachining existed in the cutting mechanism. In micromachining, the cutting mechanism was explained as the material shearing with respect to the tooltip, and it leads to chip formation. On the other hand, micromachining depends on more complex mechanisms that are relevant with the workpiece size effect. The size effect meant as the influence due to the small ratio of the cutting depth to the tool edge radius than the material acts as homogeneous and isotropic in nature. Von Turkovich and Black [13] accompanied research related to orthogonal micromachining. It was observed that the formation of chips for aluminum and copper crystals with respect to the depths of the cut had range starting from 1 to 100 μm, specifically at low cutting speed.

In another work, Nouraei et al. [14] worked on the models related to abrasive slurry jet micromachining (ASJM). In ASJM, high-pressure water was accelerated by the pump to flush out the suspended abrasive particles such as garnet or aluminum oxide (Al_2O_3). By mechanical erosion, the removal of the material takes place while in the presence of a small quantity of heat. In ASJM, high-velocity slurry jet has high machining ability that leads as a fabrication of small sizes of holes and slots [15]. Moreover, the operating parameters for the abrasive jet machining such as flow rate or pressure of slurry jet flow rate, concentration level, impact angle, and traverse speeds provided control over the erosion rate for machining various depth or width components. The plunger-type water pump is used in the process. In this system, a jet of the abrasive mixture was mixed with the high-pressure water by plunger water pump before going into exit orifice.

5.2.2 Minimum Chip Thickness and Specific Cutting Energy

In the micromachining process, both isotropic and anisotropic cutting operations were extensively inspired by the ratio of the depth of cut to the effective tool cutting edge radius. The magnitude of the chip thickness was in the same direction of the edge radius of the cutting tool. Hence, a slight variation in the depth of cut considerably affects the cutting process. The depth of cut to the effective tool cutting edge radius ratio is explained as the active material removal mechanism such as cutting, slipping, or plowing, and also this ratio is represented as the quality machining and surface roughness. Figure 5.2 represents the chip formation, sliding, and plowing in orthogonal metal cutting. Figure 5.3 illustrates the SEM photographs of typical chips formed by orthogonal flying cutting at a cutting speed of 754 m/min.

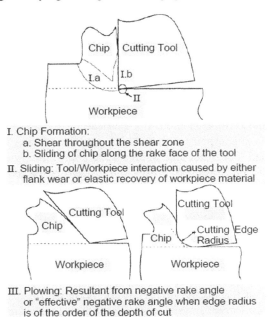

FIGURE 5.2 Chip formation, sliding, and plowing in orthogonal metal cutting [1,18].

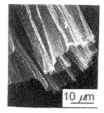 

(a) Depth of cut: 5 nm   (b) Depth of cut: 57 nm

FIGURE 5.3 SEM photographs of typical chips formed by orthogonal flying cutting (cutting speed 754 m/min) [1].

It was found that in case of minimum chip thickness, no chip will form below this minimum limit; similarly; below the minimum depth of cut, no material removal will occur. It was concluded that the identification of the required minimum chip thickness was required to make sure the accurate and precise cutting and simultaneously avoid sliding and plowing of the cutting tools [1,16]. In this work, Lucca and Seo [17] found the different effects of the geometry of single crystal diamond tool edge on the specific energy while using workpiece of Te-Cu material in the ultraprecision orthogonal fly cutting operation. Experimentation-based results and effects were investigated. The nominal rake angle had an excessive impact on the specific energy when the depth of cut was lesser than the profile of tool edge. Also the rake angle effects the specific energy, when the depth of cut approaches the size of edge contour. It was examined that the total specific energy was significantly increased at small depths of cut in micromachining of copper. Hence, the specific energy was closely related to the minimum chip thickness and would be another indicator for cutting mechanism changes and process control [18].

5.2.3 Ductile Mode Machining

In the research of Zhong et al. [19], the ductile or partial ductile modes of machining for brittle materials were considered. For fabricating the ultraprecise mirrors and an optical lens, the single-point diamond tool was used during turning operation. However, a diamond tool is not able to create the nonsymmetry shape and size for the brittle components. Nowadays, lenses and mirrors-based industries performed different shaping operation such as grinding, polishing, and lapping [20,21]. Because of the high hardness property of the ceramic materials such as silicon carbide, aluminum oxide, and zirconium oxide, to get the dimensional accuracy, the diamond grinding wheel was to get precise finishing workpiece. Metallurgical and manufacturing experts put the focus on to machine brittle materials with a lesser surface roughness or surface damage. In traditional grinding, the fractured surfaces of the workpiece required lapping and polishing operation usually for brittle materials. However, the idea of the ductile or partial ductile mode of machining for brittle materials resulted as a large range of applications on materials such as glass, silicon, germanium, and silicon carbide [22].

According to Bifano et al. [23], manufacturing of brittle material in the form of microdimensions was a brainstorming task. The fabrication of brittle material such as ceramics, optical glasses, refractory material, and so on was not precise and economical machining by conventional machining processes because of continuous variation in the brittle material geometry as well as with higher MRR. It was found that a large amount of surface roughness, as well as subsurface cracking, would be generate while machining of brittle material under traditional machining processes at a high rate of depths of cut. To get rid of that problem, machining in a ductile mode under a low depth of cuts had been proposed by many researchers. When machining occurs under a critical depth of cut to fabricate brittle materials in the form of ductile mode, it induced least surface crack with high surface finish components [1].

According to Ueda et al. [24], several ceramic materials such as ZrO_2, WC-Co, Al_2O_3, and SiC behave as a ductile materials when fabrication of ceramic materials under the ductile mode of machining declined in the depth of cut. On the other hand, if cutting speed increased, materials such as Si_3N_4 would also behave as ductile materials. It was because of high fracture toughness property of the material. Another researcher found that any brittle or hard material would be machined as a ductile fashion while modifying the depth of cut to a suitably small range [25].

5.2.4 EDGES AND SURFACE FINISH

In conventional machining, limitations such as poor edge and surface finishing, surface defects, and burrs would be the brainstorming issues. However, various process optimization and postprocesses were implemented to reduce its effects. In an article, researcher had worked in the reduction of microlevel chipping as well as cracking at the hole exit of drilling ceramic workpiece by rotary ultrasonic machining (RUM). Here, scanning electron microscope and optical microscope were employed to evaluate the exit [26]. The machining parameters were spindle speed of 1000 rpm, the feed rate of 7 μm/s, and ultrasonic power of 25%, respectively. Figure 5.4 shows the microscopic view of hole exit to visualize the chipping. Figure 5.5 describes the sectioned hole at constant drilling parameters: (i) ultrasonic machining (USM) and (ii) RUM.

It was notified that the exit chipping was apparently reduced while deploying RUM. It was concluded that the estimated response of exit crack was 24.558 μm and the experimental results were calculated as 25.378 μm [26].

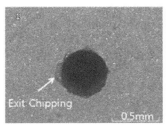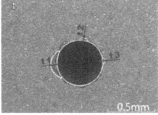

FIGURE 5.4 Microscopic view of hole exit to visualize the chipping [26].

 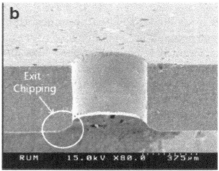

FIGURE 5.5 Description of the sectioned hole at constant drilling parameters: (a) ultrasonic machining, (b) rotary ultrasonic machining [26].

Owing to ultrasonic vibration of the tool, the trajectory motion of the diamond abrasive is sinusoidal, which leads to enhancement of hole quality [27]. During the RUM process, the brittle material was persisting tensile and shear stresses that were created by normal force [28]. In a study of RUM, the grinding process is carried out on optical BK7 and K9 glass samples. Effect of ultrasonic power on cutting forces, edge chipping, and torque were considered. It was found that at higher ultrasonic power rate, the cutting force, as well as torque, is reduced. A similar effect is also noticed in case of edge chipping. Least size of edge chipping is noticed between 40% and 80% of ultrasonic power [29]. A past researcher worked on the edge chipping reduction by making a hole on quartz glass by rotary ultrasonic drilling (RUD) process. It was stated that the reduction in cutting force leads to the crack size reduction. It was also mentioned that the cracks induced by RUD are smaller in size than induced by conventional drilling (CD) [30]. Lv et al. demonstrated a high-frequency vibration-based mechanism on hole entrance chipping in RUD process of BK7 glass. It was mentioned that during the RUD process, scale-like cracking appeared nearby hole which is lesser in amount than the defect occurred during CD [31]. Babbar et al. worked on milling of C/SiC composites that were fabricated by needling technique and chemical vapor infiltration to get good quality surface [32]. The surface quality concern has also played an important role for neurosurgical bone grinding [33,34]. Another work concentrated on the biocompatible PLA scaffold specimen for bioactivity investigation has been taken place [35]. Another technique, i.e., magnetic abrasive finishing has been used for surface finishing on flat brass workpiece [36,37]. Some other work has shown influence of metallurgical and mechanical properties, i.e., microhardness and corrosion behavior on AISI 304 SS welds to achieve best quality surface [38,39]. Hybrid activated flux has been used with tungsten inert gas welding process to refine the workpiece quality [40].

In the case of micromachining, researchers were required to do more work in this research area because of the geometrical structure limitations or characteristics that are more critical, which could not be sought out by micromachining. In medical applications, different working materials such as NiTi and surgical implants-based materials have been manufactured by micromilling technique. These materials

Miniaturization—A Boon or Curse 119

have properties such as ductility and work hardening while fabrication. It leads to high burr formation and adhesion. Due to ductility nature of the material, it leads to adverse chip formation, with long size and continuously twisted chips. As far as the microrange, the chips interference between the chip formation and the tool tip generated poor surface finish with burrs for final workpiece [41]. In another research, an experimental work performed on the microslot milling especially on copper and aluminum, which got several standard burr types depending on location and work geometry. These burrs were similar as produced in micromachining under different cutting parameters and mechanism [42]. Sharma et al. [43] have tried to eliminate the chipping using multishaped abrasive tools while creating blind holes on float glass specimen. The list of tools that were used are pin-pointed abrasive tools, hollow abrasive coated tools, flat cylindrical tools, and concave tools. Whole experiment work has been carried out by CD and RUD processes. The chipping mechanism for each multishaped abrasive tool has been reported. The final results revealed that RUD process has possessed smallest size of chipping as compared with CD. The concave circular tool was recommended as the best tool to get least chipping distance, i.e., 0.1145 mm. In another multishaped tool study, it was investigated that concave circular tool has generated least amount of tool wear. The minimum weight loss is 4.92% after CD and 1.96% after RUD using concave circular tool [44]. Sharma et al. [45] investigated the surface roughness of the float glass after RUD. The parametric optimization technique was used to achieve the least surface roughness value, i.e., 1.09 μm. The best parametric combination was noticed as feed rate (6 mm/min), spindle rotation speed (5000 rpm), and amplitude (20 μm).

5.2.5 Workpiece and Design Issues

In this research study [46], the mass production of the microcomponents is not suitable because of less rate of productivity. Microdrilling, micromilling, and microturning operation could be able to fabricate 3D concave and convex shape microcomponents, particularly by single-point diamond tool. However, to manufacture the high-quality and reliable products, the micromachining techniques would fascinate the proper tooling for the micromass production techniques such as microdiecasting, micromolding, and microfarming. It was found that the diamond tool could be the best tool to fabricate the precise as well as mold life span under micromolding process. But diamond would not be found suitable for hard ferrous material and tool steels because of its chemical instability while coming in contact with ferrous material.

In another research, Schaller et al. [47,48] presented that tungsten carbide–based tool would manufacture the martensitic steel with a width of 220–420 μm and depth of 200–330 μm. Also, these molds would fabricate metal along with plastic and ceramic.

5.2.6 Machines, Tools, and Systems for Micromachining

The development happened in previous two eras leads to an enhancement in capabilities, accuracy, and stiffness for micromachining processes over conventional machining processes. These microprocesses were widely used in the manufacturing of optical lenses as well as for surgical instruments. Some other benefits of ultraprecision machining over conventional processes would be compact space, energy, resources, reconfiguration easiness, and effortless [1]. Single-crystal diamond material has wear resistance, ease of operation, and hardness property; that is why, it was recommended for microcutting and microgrinding. The diamond tool was especially befitted to nonferrous materials, as the diamond was highly affinity to iron, such as brass, aluminum, copper, and nickel [49,50]. In another process, the focused ion beam technique has also considered manufacturing microtools. Various microgrooving, i.e., of 13 μm diameter, and microthreading cutting tools of high-speed steel and tungsten carbide are fabricated while using focused ion beam technique. With dimensional accuracy, there are also some vibration-related problems in the microtools while rotating at high speed. In the research work, Huang et al. [51] analyzed the dynamic faces of the microdrilling process. It was investigated that the natural frequency of a microdrill declined as the thrust force rises. Stiffness property also plays a vital role during manufacturing microdimensional parts. In another study, pyramidal microtools were fabricated with tip size of 2 μm to get fine surface quality and machining accuracy during micromachining of metal sheet. Also, the mechanical strength of microtools was improved. The final outcome shows that the high-quality holes have been achieved [52]. Figure 5.6 shows the five-axis control ultraprecision machining center. Figure 5.7 (a) and (b) represents the images of ultraprecise microtools and large aspect ratio microtools under optimum machining conditions.

5.2.7 CUTTING FLUID

The fundamentals of lubrication and cutting fluid used were discussed during micromachining [41]. It was found that proper quantified supply of lubrication is the essential factor in micromachining. Two important factors of lubricants are primarily, the flow pressure rate of lubricants during machining, which could affect the behavior of the cutting tool, and secondarily, after considering the diminished rate of flow

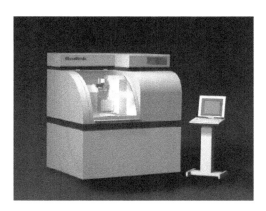

FIGURE 5.6 Commercial five-axis control ultraprecision machining center [1].

Miniaturization—A Boon or Curse

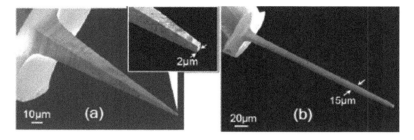

FIGURE 5.7 Overview of produced microtools under optimum machining conditions: (a) ultraprecise tool, (b) extremely large aspect ratio microtools [52].

or appropriate flow rate, exclusion of leftover cutting fluid would be a brainstorming issue during micromachining. Therefore, every machinist or engineer should exceptionally take care of courses such as cooling, transporting chips, optimum lubrication, and debris through micromachining. There were two dissimilar conditions for lubrication related to edges, and surface quality was discussed. These were MQL (minimum quantity lubrication) and dry lubrication. It was found that lubrication contains, under dry condition, burrs forming at entire trench measurement and, under minimum quantity lubrication condition, burrs forming simply at the end of the trench. Furthermore, during minimum quantity lubrication conditions, the surface side walls quality was far superior, and chip adhesion on the surface of the tool surface was far inferior. Some investigations were also carried out on the appropriate combination of nozzle distance and types of supply method, for example, continuous and intermittent type to get the best micromachining effects. Observably, MQL performs much better in micromilling processes. Various unconventional fluids such as dry ice (CO_2) were widely used in micromachining, especially for machining nickel–titanium (NiTi) material. Figure 5.8 illustrates the microend milling experiments under the minimum quality lubricant and dry conditions, where NiTi shape memory alloy was used as a workpiece material.

In a study, distill water was used as a coolant while making blind holes in float glass with five different shapes of tool. It was mentioned that apart from coolant type, shape of the tool also plays a crucial role of making fine hole quality [53]. For difficult-to-machine materials, dry machining and coolant machining could affect the generation of stresses during drilling operation The stress generation could further effect the defects such as chipping and surface roughness. In a study, drilling of float glass by RUD process has been carried out with and without coolant. During drilling operation, the maximum cutting temperature occurred with coolant (water) was 50.56°C, and 62.11°C was noticed without coolant. Accordingly, it was reported that drilling with coolant created an improved hole quality. Therefore, the stress formation has been decreased, which was occurred because of rise in cutting temperature [54]. In another work, Bissacco et al. [55] explored the accuracy rate during micromachining under the effect of working fluid. In the case of ultraprecision machine tools, these tools have a position resolution equal to 1 nm with spindles up to the high speed of 100,000 rpm. It was found that on the other hand, if the spindle speed was high, then the thermal distortion could cause inaccuracy. The cutting fluid

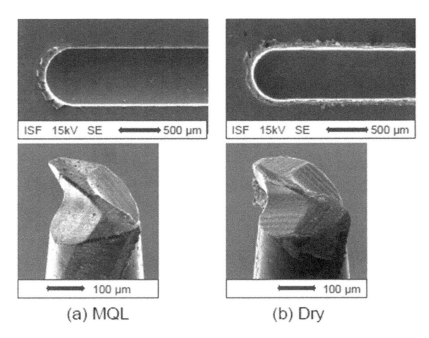

FIGURE 5.8 Microend milling experiments under (a) MQL and (b) dry conditions (work material: NiTi shape memory alloy) [1].

and the spindle temperature were also varied a lot, and it would lead to an error. At these stages, the minor offset during machining produced by thermal distortion may lead to substantial errors. In a work of rotary ultrasonic milling, C/SiC composite has been fabricated by varying material density using chemical vapor infiltration and needling method. It was noticed that as the material density is increased, the average surface roughness value reached to 0.84 μm. Moreover, it was also mentioned that the tool wear has also affected the material density [56]. In some studies, biocompatible material (PLA scaffold) and neurosurgical bone–related work have been reported [33,57].

5.2.8 Machine Components and Controls

During micromachining, advancement of machine tool components and simultaneously study of process physics and its control units were desirable to acquire the best quality and high accuracy [58]. Studies related to high-speed spindle, accurate positioning systems, jigs, and fixture devices were most essential for the optimum micromachining. Development related to ultraprecision positioning system was carried out while using a stepper motor and ball screw with 1 nm positional resolution with piezoelectric actuators.

Mizumoto et al. [59] established a mechanism based upon the trimode ultraprecision positioning along with twisting roller drive and also the aerostatic guideway with an active integral restrictor. The position-based resolution range of this instrument is

Miniaturization—A Boon or Curse

25 pm. Holmes et al. [60] developed a magnetic-bearing motion control stage with a positional resolution accuracy of 0.1 nm.

5.2.9 Metrology in Micromachining

Nowadays, it is understood that the micromachining techniques have the capacity to manufacture microsize components of very precise dimension stability. To improve the manufacturing and control, the smaller feature components are required to get the precise quality. Monitoring by sensors provides treasured information about the micromachining practice that helps the double purpose of the monitor such as quality and process control and delivering the fully-automated manufacturing environment by using sensors [61]. In this research work, various micromachining-related measurement methods were considered. These methods were facilitated to measure the dynamics of displacement as well as force, properties of material, and shape and size during machining at the microlevel. As the dimensional ranges of the parts get diminished, the resolution techniques play a vital role to measure and compute it. Alongside, market demand rises to get an accurate, precise, and efficient measurement of different applications stretching from simple geometry components to complex geometry objects such as microholes or microspheres. Commonly, micromachining-quantified instruments or devices can be classified into two separated groups. In the first group, a grouping of measurement instruments measured either height or changes in height. In this group, optical profile followers, atomic force microscopes, mechanical stylus instruments, and interference measurement devices such as laser interferometer devices could be included. Second group includes, measurement of the distance between edges of an article, electron microscopes and optical profilometer, it included such as mechanical, magnetic, optical and capacitive sensor along with workpiece holding table and a transducer to relocate the workpiece [62,63]. After investigation of such devices, still there are some cases such as microparts including inside features such as holes, channels, and pockets. Here, Figure 5.9 showed the level of precision and control parameters that varied with respect to sensor application.

These are those microparts or components where still there exists no technology to measure such structures. In this study, Masuzawa et al. [64] established a vibroscanning technique to quantify the inside dimensions of microholes. But this method has some constraints such as it is restricted to only conductive materials. The cause of this restriction is that it practices a sensitive electrical switch by contacting a vibrating microprobe onto the workpiece. There is an additional probe used which utilized contact by bending of the probe [65]. Miyoshi et al. [66] established a profile measurement technique. It included the inverse scattering phase retrieval method. This test could measure surface profile with submicron accuracy along with symmetric and nonsymmetric fine triangular grooves. In a work of Sharma et al. [67], coordinate measuring machine has been used to quantify the chipping measurements in horizontal and vertical directions at drilled hole exit corner of float glass. Finally, the volume of chipping is quantified accordingly.

Although to get more accuracy and precise measurements of the microparts, the coordinate measuring machine is a vital tool. Still, there are necessities of

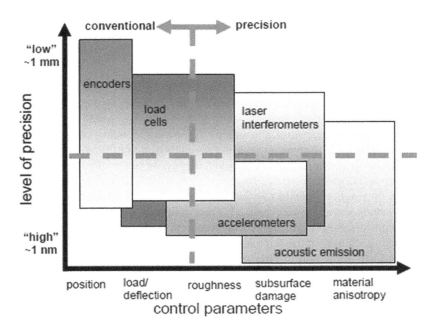

FIGURE 5.9 Sensor application versus level of precision and control parameters [61].

micromachining metrology instruments or devices to measure the various shape and sizes of microparts, which is needed to be carried out in future.

EDM finds good application in micromachining of biomedical components. Biomedical components are usually very hard and difficult to machine by conventional methods. Complicated shapes and very precise accuracy are required in transplanting bone-related plates to human body. EDM has good scope in this area. Basak et al. [68] investigated machining of titanium alloy with wire EDM. By varying the flushing pressure, wire tension, and pulse on time, they have studied the effects of kerf width, MRR, and discharge gap. It is found wire is deformed after experiment and recast later with cracks and holes. Prakash et al. [69] have used powder-mixed EDM for studying its effect on a b-Ti-based implant for producing a biomimetic nanoporous bioceramic surface. It has been successfully proved by them that a natural bone-like surface structure can be produced by this method. Aliyu et al. [70] have successfully studied machining of bulk metallic glasses with hydroxyapatite-mixed electrical discharge machining (HA-EDM). The surface produced by HA-EDM is expected to facilitate higher tissue ingrowth and bone implant adhesion. Prakash et al. [71] have fabricated an alloy by depositing a nanohydroxyapatite (nHA) coating on biodegradable Mg-Zn-Mn alloy. The main purpose of coating is to reduce the corrosion of implants. It is proved that such coating helps in reducing corrosion at a considerable level. Prakash et al. [72] have used powder-mixed EDM and nondominated sorting genetic algorithm for fabricating biocompatible surface on β-phase Ti alloy. After optimizing and confirmation, the result shows 184% increase in microhardness. They have used response surface methodology for predicting the results. EDM, Powder mixed electrical discharge machining (PMED), and W-EDM (wire-EDM)

Miniaturization—A Boon or Curse

have been found potential to enhance the Mg- and Ti-based implants for biomedical applications [70,73–100].

5.3 CONCLUSION AND BRIEF DISCUSSION

Some of the decisive conclusions are summarized as in the following:

- Some of the major applications of micromachining are ICs, small surgical tools, and fuel injection nozzles, which makes the micromachining-based study as a trending area of research.
- The diamond tool was especially befitted to nonferrous materials. Since the diamond was highly affinity to iron, such as brass, aluminum, copper, and nickel. Many researchers are currently focused on such diamond-coated abrasive tools for machining purposes.
- Multishaped abrasive tools were used while creating blind holes on float glass specimen to reduce the chipping at hole entrance. Pin-pointed abrasive tools, hollow abrasive-coated tools, flat cylindrical tools, and concave tools were used during experimentation. The concave circular tool was recommended as the best tool to get least chipping distance, i.e., 0.1145 mm.
- Several ceramic materials such as zirconium dioxide, tungsten carbide, aluminum oxide, and silicon carbide (SiC) behave as a ductile material when machining occurs under ductile mode of machining as declined in the depth of cut.
- As compared with CD process, RUM technique is recommended to achieve chipping-free machining with least tool wear.
- MQL performs much better in micromilling processes. Various unconventional fluids such as dry ice (CO_2) are widely used in micromachining, especially for machining nickel–titanium-based (NiTi) materials.
- To get more accuracy and precise measurements of the microparts, the coordinate measuring machine is a vital tool.

5.4 FUTURE SCOPE

Several key points of this research study, which encouraged further future investigation and research in the area of micromachining, are as follows:

- It was found that the diamond tool could be the best tool to fabricate the precise as well as mold life span under micromolding process. But diamond would not be found suitable for hard ferrous material and tool steels because of its chemical instability while coming in contact with ferrous material.
- To get optimized dimensional accuracy, there are also some vibration-related problems in the microtools while rotating at high speed.
- Considering the diminished rate of flow or appropriate flow rate, exclusion of leftover cutting fluid is a brainstorming issue during micromachining. Therefore, every machinist or engineer should exceptionally take care of courses such as cooling, transporting chips, optimum lubrication, and debris through micromachining.

- Still, there are necessities of micromachining metrology instruments or devices to measure the various shape and sizes of microparts including inside features such as holes, channels, and pockets.
- It was discussed that the more research work is still required in the area of electrochemical micromachining to do a deep study of its parametric optimization, local MRR, process mechanism, current density, and so on.
- Furthermore, there would be a plethora of research studies and advancements in the field of Engineering Metrology and Measurements (EMM) that would be required in the various electronics and precision industry and exemplified as ultraprecision microfabrication, surface finishing of print bands, deburring, and 3D micromachining.
- Area related to the metrology of the microparts is a brainstorming research area which is needed to be carried out in future.

REFERENCES

1. Dornfeld D, Min S, Takeuchi Y (2006) Recent advances in mechanical micromachining. *Annals of CIRP* 55 (2): 745–768.
2. Taniguchi N (1983) Current status in and future trends of ultra-precision machining and ultrafine materials processing. *CIRP Annals* 32 (2): 573–582.
3. Ghosh A, Mallik AK (1986) *Manufacturing Science (Book)*. Ellis Horwood ltd. 1st Edition: 1–707, ISBN- 0470203129, 9780470203125.
4. Rajput RK (2007) *A Textbook of Manufacturing Technology: Manufacturing Processes*. Firewell Media: 1–899, ISBN- 9788131802441.
5. Kalpakjain S, Schmid S (2010) *Manufacturing Engineering and Technology*. Prentice Hall, Pearson. 7th Edition: 1–1216, ISBN- 10: 0133128741.
6. Alting L, Kimura F, Hansen HN, Bissacco G (2003) Micro engineering. *CIRP Annals* 52 (2): 635–657.
7. Matthias HO, Gunnar M, Zinecker M, Andre M, Andreas S (2012) Micro machining with continuous electrolytic free jet. *Precision Engineering* 36 (4): 612–619.
8. Kunieda M, Mizugai K, Watanabe S, Shibuya N, Iwamoto N (2011) Electrochemical micromachining using flat electrolyte jet. *CIRP Annals-Manufacturing Technology* 60 (1): 251–254.
9. Jian L, Juan L, Chengying X (2014) Interaction of the cutting tools and the ceramic reinforced metal matrix composites during micro-machining: a review. *CIRP Annals* 7 (2): 55–70.
10. Goh CS (2007) Characterization of high performance Mg/MgO nanocomposites. *Journal of Composite Materials* 41 (19): 2325–2335.
11. Yan H, Wan J, Nie Q (2013) Wear behavior of SiCp-reinforced magnesium matrix composites. *Wear* 255 (1–6): 629–637.
12. Aist E, Elsaber M, Hort N, Limberg W (2006) Machining of hybrid reinforced Mg-MMCs using abrasive water jetting, 7th Magnesium Technology Symposium: 345–348.
13. Von Turkovich BF, Black JT (1970) Micro-machining of copper and aluminum crystals. *Transactions of the ASME* 92 (1): 130–134.
14. Nouraei H, Kowsari K, Spelt JK, Papini M (2014) Surface evolution models for abrasive slurry jet micro-machining of channels and holes in glass. *Wear* 309 (1–2): 65–73.
15. Nouraei H, Wodoslawsky A, Spelt JK, Papini M (2011) Micro-machining using an Abrasive Slurry Jet. *Wear of Materials, International Conference*, USA.

16. Ikawa N, Shimada S, Tanaka H (1992) Minimum thickness of cut in micromachining. *Nanotechnology* 3 (1): 6–9.
17. Lucca DA, Seo YW (1993) Effect of tool edge geometry on energy dissipation in ultra-precision machining. *CIRP Annals* 42 (1): 83–88.
18. Lucca DA, Rhorer RL, Komanduri R (1991) Energy dissipation in the ultra-precision machining of copper. *CIRP Annals* 40 (1): 69–72.
19. Zhong ZW (2003) Ductile or partial ductile mode machining of brittle materials. *The International Journal of Advanced Manufacturing Technology* 21: 579–585.
20. Balson P, Pung R (1991) *Diamond Wheel Selection Criteria for Grinding Advanced Engineering Ceramics.* Society of Manufacturing Engineers: 1–20.
21. Malkin S, Ritter J E (1989) Grinding mechanisms and strength degradation for ceramics. *Journal of Engineering for Industry* 111 (2): 167–174.
22. Blake PN, Scattergood RO (1990) Ductile regime machining of germanium and silicon. *Journal of American Ceramic Society* 73 (4): 949–957.
23. Bifano TG, Dow TA, Scattergood RO (1991) Ductile regime grinding: A new Technology for machining brittle materials. *Journal of Engineering for Industry, Transactions of the ASME* 113 (2): 184–189.
24. Ueda K, Sugita T, Hiraga H (1991) J-Integral approach to material removal mechanisms in microcutting of ceramics. *CIRP Annals* 40 (1): 61–64.
25. Shimada S, Ikawa N, Inamura T, Takezawa N, Ohmori H, Sata T (1995) Brittle ductile transition phenomena in microindentation and micromachining. *CIRP Annals* 44 (1): 523–526.
26. Liu JW, Baek DK, Ko TJ (2014) Chipping minimization in drilling ceramic materials with rotary ultrasonic machining. *The International Journal of Advanced Manufacturing Technology* 72: 1527–1535.
27. Wang J, Zhang J, Feng P, Guo P (2018) Damage formation and suppression in rotary ultrasonic machining of hard and brittle materials: A critical review. *Ceramics International* 44 (2): 1227–1239.
28. Singh, RP, Singhal S (2016) Rotary ultrasonic machining: A review. *Materials and Manufacturing Processes* 31 (14):1795–1824.
29. Hu Y, Wang H, Ning F, Li Y, Cong W (2017) Surface grinding of optical bk7/k9 glass using Rotary ultrasonic Machining: An experimental study. *Proceedings of the ASME-12th International Manufacturing Science and Engineering Conference*, American Society of Mechanical Engineers Digital Collection: 1–7.
30. Wang J, Zha H, Feng P, Zhang J (2016) On the mechanism of edge chipping reduction in rotary ultrasonic drilling: A novel experimental method. *Precision Engineering* 44: 231–235.
31. Lv D, Jhang Y, Peng Y (2016) High-frequency vibration effects on hole entrance chipping in rotary ultrasonic drilling of BK7 glass. *Ultrasonics* 72: 47–56.
32. Babbar A, Sharma A, Jain V, Jain AK (2019) Rotary ultrasonic milling of C/SiC composites fabricated using chemical vapor infiltration and needling technique. *Materials Research Express* 6 (8): 085607.
33. Babbar A, Jain V, Gupta D (2019) Neurosurgical bone grinding. In *Biomanufacturing*. Springer International Publishing, Springer, Cham: 137–155.
34. Babbar A, Jain V, Gupta D (2019) Thermogenesis mitigation using ultrasonic actuation during bone grinding: A hybrid approach using CEM43°C and Arrhenius model. *Journal of the Brazilian Society of Mechanical Sciences and Engineering* 41 (10): 401.
35. Babbar A, Singh P, Farwaha HS (2017) Regression model and optimization of magnetic abrasive finishing of flat brass plate. *Indian Journal of Science and Technology* 10: 1–7.
36. Babbar A, Singh P (2017) Parametric study of magnetic abrasive finishing of UNS C26000 flat brass plate. *International Journal of Advanced Multidisciplinary Research* 9 (2): 83–89.

37. Kumar M, Babbar A, Sharma A, Shahi AS (2019) Effect of post weld thermal aging (PWTA) sensitization on micro-hardness and corrosion behavior of AISI 304 weld joints. *IOP Conference Series: Journal of Physics: Conference Series*, IOP Publishing, 1240: 012078.
38. Sharma A, Kumar M, Shahi AS (2018) A sensitization studies on the metallurgical and corrosion behavior of AISI 304 SS welds. In *Advances in Manufacturing Processes, Lect. Notes Mechanical Engineering*. Chapter-17: 257–265. DOI: 10.1007/978-981-13-1724-8_25.
39. Babbar A, Kumar A, Jain V, Gupta D (2019) Enhancement of activated tungsten inert gas (A-TIG) welding using multi-component TiO2-SiO2-Al2O3 hybrid flux. *Measurement* 148: 106912.
40. Singh D, Babbar A, Jain V, Gupta D, Saxena S, Dwibedi V (2019) Synthesis, characterization, and bioactivity investigation of biomimetic biodegradable PLA scaffold fabricated by fused filament fabrication process. *Journal of the Brazilian Society of Mechanical Sciences and Engineering* 41 (3):121.
41. Weinert K, Kahnis P, Petzoldt V, Peters C (2005) Micro-Milling of steel and NiTi SMA. *55th CIRP General Assembly, STC-C Section Meeting Presentation File*, Turkey.
42. Lee K, Dornfeld DA (2002) An experimental study on burr formation in micro milling aluminum and copper. *Transactions of the NAMRI/SME* 30: 1–8.
43. Sharma A, Jain V, Gupta D (2019) Comparative analysis of chipping mechanics of float glass during rotary ultrasonic drilling and conventional drilling: For multi-shaped tools. *Machining Science and Technology*. DOI: 101080/10910344.2019.1575402.
44. Sharma A, Jain V, Gupta D (2019) Multi-shaped tool wear study during rotary ultrasonic drilling and conventional drilling for amorphous solid. Proceedings of the Institution of Mechanical Engineers, Part E: *Journal of Process Mechanical Engineering* 233: 551–560.
45. Sharma A, Jain V, Gupta D (2018) Enhancement of surface roughness for brittle material during rotary ultrasonic machining. In *MATEC Web of Conferences*. EDP Sciences 249: 01006. doi: https://doi.org/10.1051/matecconf/201824901006.
46. Klocke F, Weck M, Fischer S, Ozmeral H, Schroeter RB, Zamel S (1996) Ultraprecision machining and manufacturing of micro components. *Diamond and Related Materials* 3: 172–177.
47. Schaller T, Heckele M, Ruprecht R, Schubert K (1999) Micro fabrication of a mold insert made of hardened steel and first molding results. *Proceedings of the ASPE* 20: 224–227.
48. Schaller T, Mayer J, Schubert K (1999) Approach to a micro structured mold made of steel. *Proceedings of the EUSPEN* 1: 238–241.
49. Brinksmeier E, Malz R, Riemer O (1996) Micromachining ductile and brittle materials in optical quality. *VDI Berichte* 221–229.
50. Weck M, Fischer S, Vos M (1997) Fabrication of micro components using ultraprecision machine tools. *Nanotechnology* 8 (3): 145–148.
51. Huang BW (2004) The drilling vibration behavior of a twisted Micro drill. *Journal of Manufacturing Science and Engineering. Transactions of the ASME* 126 (4): 719–726.
52. Ohmori H, Katahira K, Uehara Y, Watanabe Y, Lin W (2003) Improvement of mechanical strength of micro tools by controlling surface characteristics. *CIRP Annals* 52 (1): 467–470.
53. Sharma A, Jain V, Gupta D. (2018) Tool wear analysis while creating blind holes on float glass using conventional drilling: A multi-shaped tools study. In *Advances in Manufacturing Processes. Proceedings of ICEMMM, Lect. Notes Mechanical Engineering*, Chapter-17. Springer, Singapore: 175–183.
54. Sharma A, Jain V, Gupta D. (2019) Experimental investigation of cutting temperature during drilling of Float glass specimen. *3rd International Conference on Aerospace, Mechanical and Mechatronic Engineering (CAMME)*: 1–6 (In press).

55. Bissacco G, Hansen HN, Chiffre DL (2005) Micro-milling of hardened tool steel for mold making applications. *Journal of Materials Processing Technology* 167 (2–3): 201–207.
56. Otsuka J, Hata S, Shimokohbe A, Koshimizu S (1998) Development of ultra-precision table for ductile mode cutting. *Journal of the Japan Society for Precision Engineering* 64 (4): 546–551.
57. Singh D, Babbar A, Jain V, Gupta D, Saxena S, Dwibedi V (2019) Synthesis, characterization, and bioactivity investigation of biomimetic biodegradable PLA scaffold fabricated by fused filament fabrication process. *Journal of the Brazilian Society of Mechanical Sciences and Engineering* 41 (3): 121.
58. Otsuka J, Ichikawa S, Masuda T, Suzuki K (2005) Development of a small ultra-precision positioning device with 5 nm resolution. *Measurement Science and Technology* 16 (11): 2186–2192.
59. Mizumoto H, Yabuta Y, Arii S, Tazoe Y, Kami Y (2005) A Picometer positioning system using active aerostatic guide way. *Proceedings of the International Conference on Leading Edge Manufacturing in 21st Century*, The Japan Society of Mechanical Engineers: 1009–1014.
60. Holmes M, Trumpet D, Hocken R (1995) Atomic scale precision motion control stage. *CIRP Annals* 44 (1): 455–460.
61. Lee DE, Hwanga I, Valente CMO, Oliveira JFG, Dornfeld DA (2000) Precision manufacturing process monitoring with acoustic emission. *International Journal of Machine tools & Manufacture* 46: 176–188.
62. Umeda A (1996) Review on the importance of measurement technique in micro machine technology. *Proceedings of SPIE-The International Society for Optical Engineering* 2880: 26–38.
63. McGeough JA (2002) *Micromachining of Engineering Materials*. Marcel Dekker Inc., New York.
64. Masuzawa T, Hamasaki Y, Fujino M (1993) Vibro-scanning method for non-destructive measurement of small holes. *CIRP Annals* 42 (1): 589–592.
65. Kim B, Masuzawa T, Bourouina T (1999) Vibroscanning method for the measurement of micro hole profiles. *Measurement Science and Technology* 10 (8): 697–705.
66. Miyoshi T, Takaya Y, Saito K, (1996) Micro-machined profile measurement by means of optical inverse scattering phase method. *CIRP Annals* 45 (1):497–500.
67. Sharma A, Jain V, Gupta D. (2018) Characterization of chipping and tool wear during drilling of float glass using rotary ultrasonic machining. *Measurement* 128: 254–263.
68. Pradhan S, Singh S, Prakash C, Królczyk G, Pramanik A, Pruncu CI (2019) Investigation of machining characteristics of hard-to-machine Ti-6Al-4V-ELI alloy for biomedical applications. *Journal of Material Research and Technology* 8 (5): 4849–4862.
69. Basak A, Pramanik A, Prakash C (2019) Surface, kerf width and material removal rate of Ti6Al4V titanium alloy generated by wire electrical discharge machining. *Heliyon* 5: 1–17.
70. Prakash C, Kansal HK, Pabla BS, Puri S (2015) Processing and characterization of novel biomimetic nanoporous bioceramic surface on β-Ti implant by powder mixed electric discharge machining. *Journal of Materials Engineering and Performance* 24: 3622–3633, 2015.
71. Aliyu AA, Abdul-Rani AM, Ginta TL, Prakash C, Axinte E, Fua-Nizan R (2017) Fabrication of nanoporosities on metallic glass surface by hydroxyapatite mixed EDM for orthopedic application. *International Medical Device and Technology Conference, Malaysia*: 168–171.
72. Prakash C, Singh S, Pabla BS, Uddin MS (2018) Synthesis, characterization, corrosion and bioactivity investigation of nano-HA coating deposited on biodegradable Mg-Zn-Mn alloy. *Surface and Coatings Technology* 346: 9–18.

73. Prakash C, Singh S, Singh M, Verma K, Chaudhary B, Singh S (2018) Multi-objective particle swarm optimization of EDM parameters to deposit HA-coating on biodegradable Mg-alloy. *Vacuum* 158: 180–190.
74. Prakash C, Kansal HK, Pabla BS, Puri S (2016) Multi-objective optimization of powder mixed electric discharge machining parameters for fabrication of biocompatible layer on β-Ti alloy using NSGA-II coupled with Taguchi based response surface methodology. *Journal of Mechanical Science and Technology* 30 (9): 4195–4204. DOI: 10.1007/s12206-016-0831-0.
75. Prakash C, Uddin MS (2017) Surface modification of β-phase Ti implant by hydroxyapatite mixed electric discharge machining to enhance the corrosion resistance and in-vitro bioactivity. *Surface Coating and Technology* 236 (Part A): 134–145.
76. Prakash C, Singh S, Singh R, Ramakrishna S, Pabla BS, Puri S, Uddin MS (2019) *Biomanufacturing*. Springer.
77. Singh H, Singh S, Prakash C (2019) Current trends in biomaterials and biomanufacturing. In *Biomanufacturing*. Springer, Cham: 1–34.
78. Prakash C, Kansal HK, Pabla BS, Puri S, Aggarwal A (2016) Electric discharge machining – A potential choice for surface modification of metallic implants for orthopedic applications: A review. *Proceedings of the Institution of Mechanical Engineers, Part B: Journal of Engineering Manufacture* 230 (2): 331–353.
79. Prakash C, Kansal HK, Pabla BS, Puri S (2017) Experimental investigations in powder mixed electric discharge machining of Ti–35Nb–7Ta–5Zrβ-titanium alloy. *Materials and Manufacturing Processes* 32 (3):274–285.
80. Prakash C, Kansal HK, Pabla BS, Puri S (2016) Powder mixed electric discharge machining: an innovative surface modification technique to enhance fatigue performance and bioactivity of β-Ti implant for orthopedics application. *Journal of Computing and Information Science in Engineering* 16 (4), 041006.
81. Prakash C, Kansal HK, Pabla BS, Puri S (2015) Potential of powder mixed electric discharge machining to enhance the wear and tribological performance of β-Ti implant for orthopedic applications. *Journal of Nanoengineering and Nanomanufacturing* 5 (4): 261–269.
82. Prakash C, Singh S, Verma K, Sidhu SS, Singh S (2018) Synthesis and characterization of Mg-Zn-Mn-HA composite by spark plasma sintering process for orthopedic applications. *Vacuum* 155: 578–584.
83. Prakash C, Singh S, Pruncu CI, Mishra V, Królczyk G, Pimenov DY, Pramanik A (2019) Surface modification of Ti-6Al-4V alloy by electrical discharge coating process using partially sintered Ti-Nb electrode. *Materials* 12 (7): 1006.
84. Prakash C, Kansal HK, Pabla BS, Puri, S (2017) On the influence of nanoporous layer fabricated by PMEDM on β-Ti implant: biological and computational evaluation of bone-implant interface. *Materials Today: Proceedings* 4 (2): 2298–2307.
85. Prakash C, Kansal HK, Pabla BS, Puri S (2016) Effect of surface nano-porosities fabricated by powder mixed electric discharge machining on bone-implant interface: An experimental and finite element study. *Nanoscience and Nanotechnology Letters* 8 (10): 815–826.
56. Prakash C, Kansal HK, Pabla BS, Puri S (2015, December) To optimize the surface roughness and microhardness of β-Ti alloy in PMEDM process using non-dominated sorting genetic algorithm-II. In *2015 2nd International Conference on Recent Advances in Engineering & Computational Sciences (RAECS)*. IEEE: 1–6.
87. Prakash C, Kansal HK, Pabla BS, Puri S (2017) Potential of silicon powder-mixed electro spark alloying for surface modification of β-phase titanium alloy for orthopedic applications. *Materials Today: Proceedings* 4 (9): 10080–10083.
88. Pramanik A, Basak AK, Prakash C (2019) Understanding the wire electrical discharge machining of Ti6Al4V alloy. *Heliyon* 5 (4): e01473.

89. Pramanik A, Islam MN, Basak AK, Dong Y, Littlefair G, Prakash C (2019) Optimizing dimensional accuracy of titanium alloy features produced by wire electrical discharge machining. *Materials and Manufacturing Processes*, 34 (10): 1083–1090.

90. Prakash C, Singh S, Singh M, Antil P, Aliyu AAA, Abdul-Rani AM, Sidhu SS (2018) Multi-objective optimization of MWCNT mixed electric discharge machining of Al–30SiC p MMC using particle swarm optimization. *Futuristic Composites Behavior, Characterization, and Manufacturing*, Springer Nature Singapore Pte Ltd: 145–164.

91. Prakash C, Kansal HK, Pabla BS, Puri S (2016, March) Research on the formation of nano-porous biocompatible layer on Ti-6Al-4V implant by powder mixed electric discharge machining for biomedical applications. In *International Conference*, NanoSciTech.

92. Antil P, Singh S, Manna A, Prakash C (2018) Electrochemical discharge drilling of polymer matrix composites. In *Futuristic Composites*. Springer, Singapore: 223–243.

93. Antil P, Singh S, Singh S, Prakash C, Pruncu CI (2019) Metaheuristic approach in machinability evaluation of silicon carbide particle/glass fiber–reinforced polymer matrix composites during electrochemical discharge machining process. *Measurement and Control* 52 (7–8): 1167–1176.

94. Abdul-Rani AM, Ginta TL, Prakash C, Rao TVVLN, Axinte E, Ali S (2019, May) Synthesis and characterization of bioceramic oxide coating on Zr-Ti-Cu-Ni-Be BMG by electro discharge process. In *International Scientific-Technical Conference MANUFACTURING*. Springer, Cham: 518–531.

95. Malik A, Pradhan S, Mann GS, Prakash C, Singh S (2019) Subtractive versus hybrid manufacturing. *Encyclopedia of Renewable and Sustainable Materials* 5: 474–502.

96. Prakash C, Singh S, Sharma S, Singh J, Singh G, Mehta M, Mittal M, Kumar H (2020) Fabrication of low elastic modulus Ti50Nb30HA20 alloy by rapid microwave sintering technique for biomedical applications. *Materials Today: Proceedings* 21: 1713–1716.

97. Prakash C, Singh S, Ramakrishna S, Królczyk G, Le CH (2020) Microwave sintering of porous Ti–Nb-HA composite with high strength and enhanced bioactivity for implant applications. *Journal of Alloys and Compounds* 824: 153774.

98. Prakash C, Singh S (2020) On the characterization of functionally graded biomaterial primed through a novel plaster mold casting process. *Materials Science and Engineering: C* 110: 110654.

99. Prakash C, Singh S, Sharma S, Garg H, Singh J, Kumar H, Singh G (2019) Fabrication of aluminium carbon nano tube silicon carbide particles based hybrid nano-composite by spark plasma sintering. *Materials Today: Proceedings* 21: 1637–1642.

100. Prakash C, Singh S, Basak A, Królczyk G, Pramanik A, Lamberti L, Pruncu CI (2020) Processing of Ti50Nb50– xHAx composites by rapid microwave sintering technique for biomedical applications. *Journal of Materials Research and Technology* 9 (1): 242–252.

6 A Comprehensive Review on Similar and Dissimilar Metal Joints by Friction Welding

D. Saravanakumar, M. Uthayakumar, and S. Thirumalai Kumaran
Kalasalingam Academy of Research and Education

CONTENTS

6.1 Introduction 133
6.2 Friction Welding between Ferrous and Nonferrous Metal Alloys 134
6.3 Friction Welding between Ferrous Metal Alloys 135
6.4 Friction Welding between Nonferrous Metal Alloys................................. 141
6.5 Finite Element Model in Friction Welding.. 142
6.6 Conclusion 143
References... 144

6.1 INTRODUCTION

Friction welding (FW) is a planet-friendly process because it will not produce any kind of fumes and smokes and health hazards are minimal (Figure 6.1). FW is a kind of solid-state metal joining process, which is used to fuse the similar or dissimilar metal alloys with the help of various parameters such as frictional force, friction time, friction speed, forge pressure, and forge time [1]. Rotary friction welding (RFW) is a simple form of FW process. In this setup, one component is static form and another component is rotational form which is forced to rub against the static component for fusing. Continuous friction welding (CFW) and inertia friction (IFW) welding are two variants of RFW. In CFW, the dynamic part is coupled with motor-driven unit which gives constant rotational speed throughout the welding process. In IFW, the dynamic part is coupled with flywheel which is used to vary the rotational speed.

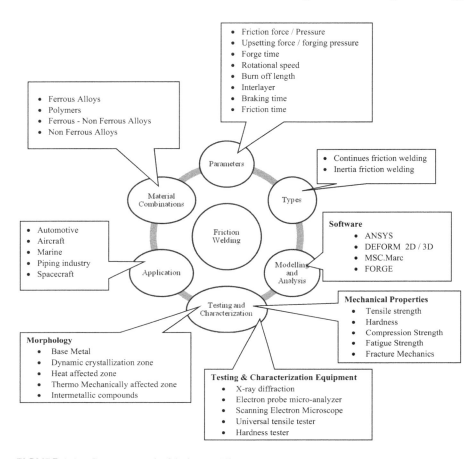

FIGURE 6.1 Components in friction welding.

6.2 FRICTION WELDING BETWEEN FERROUS AND NONFERROUS METAL ALLOYS

Kimura et al. [2] investigated the morphology, joint efficiency, natural aging treatment, and mechanical properties such as tensile strength, hardness, and bend ductility of friction-welded joints among AA6063 and AISI304 metal alloys. The experimental condition was carried at a friction speed of 1650 rpm, 30 MPa friction force, friction time from 0.04 to 5.0 s, forge pressure of 30–240 MPa, and time of 6.0 s. The result shows that, to obtain maximum joint efficiency, the joint should be sustained with high forge pressure and temperature. Radoslaw Winiczenko et al. [3] examined the mechanical properties and microstructure analysis of friction-welded joint between AA5454 aluminum alloy and W-Ni-Fe (7:3 nickel to iron ratio). By making the different samples, the range of friction pressure and friction time varied. From this study, it is observed that average tensile strength of 234 MPa is attained at a friction time of 3.5 s and friction pressure of 40 MPa. The microstructure of friction-welded joint indicated that fine grains

Similar and Dissimilar Metal Joints by FW

due to dynamic recrystallization of alloys and coarse grains were presented at periphery of aluminum alloy side. Dey et al. [4] have studied corrosion behavior, mechanical properties, and morphological analysis in dissimilar joint of Ti–304L stainless steel (SS). In addition of postweld heat treatment (PWHT), bend ductility was increased, and residual stress was eliminated from the joint interface. They found that the average corrosion rate at dissimilar joint interface was 10 mpy (mils penetration per year) under boiling nitric acid environment. The FW AA6061 to AISI4340 combination was inspected with and without silver interlayer. The interdiffusion of silver as an interlayer showed better results in tensile strength and ductility at welded region [5]. Ti-6Al-4V–SS304L joint was made with oxygen-free copper interlayer. They evaluated the momentous of copper as an interlayer. Moreover, the presence of copper between the welded joint showed better tensile strength compared with noninterlayer joint [6]. Ahmet Hascalik et al. [7] prepared weld joint with Al_2O_3 particle–reinforced Al6061 composite and SAE1020 and compared oxide-fragmented region in aluminum composite side and plastic-deformed region in steel side with base metals. The results indicated that the increases of plastic deformation are directly proportional to the size and volume fraction of reinforced particles. Powder metallurgical (PM) steel and wrought copper dissimilar joint was achieved in continuous drive FW by Jayabharath et al. [8]. They concluded that low-density metals are with lower process parameters and high-density metals are with higher process parameters. Vairamani et al. [9] fabricated AISI 304/Cu alloy joint and investigated with optimized parameters to achieve minimum hardness at weld interface region with an aid of response surface methodology. They had formulated experimental design matrix to analyze the effect of parameter variance. Hazman Seli et al. [10] performed some characterization and mathematical thermal modeling in alumina and mild steel friction-welded joint with Al1100 as an interlayer. They observed maximum bending strength at 180 MPa of friction pressure and 20 s of friction time. Table 6.1 shows the collective literatures of friction welding between ferrous and nonferrous metal alloys.

6.3   FRICTION WELDING BETWEEN FERROUS METAL ALLOYS

Javed Akram et al. analyzed the creep behavior of P91/AISI304 single-interlayer friction-welded joint through Inconel 600 and three-interlayer friction-welded joint with Inconel 600, Inconel 625, and Inconel 800H. Proved three Inconel interlayer friction-welded sample exhibited better rupture life and noted transgranular mode of fracture in all samples [11]. Furkan Sarsilmaz et al. studied joinability, microstructure, and mechanical properties of Armor 500 and duplex AISI2205 steels. The highest tensile strength of 1020 MPa was achieved at 8 s friction timing, 80 MPa friction pressure, 120 MPa upsetting pressure, and 1800 rpm [12]. Jeswin Alphy James et al. [13] worked on with and without nickel as an interlayer in between SS304 and AISI1040 dissimilar friction-welded steels. Result shows that maximum tensile strength is achieved with presence of nickel interlayer and hardness is decreased due to precipitation of chromium carbide.

136 Advanced Manufacturing and Processing Technology

TABLE 6.1

Collective literatures of friction welding between ferrous and nonferrous metal alloys.

Ref	Author	Specimen Dimension, Diameter (D), Length (L)	Friction Force/ Pressure	Friction Time	Upsetting force/Forge Pressure	Upsetting/ Forge time	Rotational Speed (rpm)	Burnoff Length	Braking Time	Interlayer
[1]	Suresh	AA6061–AISI4340	D 16 mm	3 kN		6 kN		2400	2 mm		Ag
[2]	Kimura M	AA6063–AISI304	D 16 mm	30 MPa	0.04–5.0 s	30–240 MPa	6 s	1650			
[3]	Radoslaw	AA5454-W-Ni-Fe	D 20 mm, L 100 mm	40–80 MPa	0.5–9.5 s	159 MPa	5 s	1450		0.1 s	
[4]	Javed	P91–AISI304	D 25.4 mm	160 MPa		280 MPa		1500	5 mm		IN600, IN625, IN800H
[5]	Jeswin	SS304 – AISI1040	D 10 mm L 75 mm for AISI 1040 D 10 mm L 55 mm for SS304			1.570–1.884 ton		2100–2500	4–8 mm		Ni
[6]	Prasanthi	MS–Grade-2 Ti	D 10 mm L 100 mm	0.8–1 ton		1.6–2 ton		750–1500	2–3 mm		
[7]	Yanni Wei	Al–Cu	D 20 mm	19.1 MPa	2–8 s	31.8 MPa		1900			
[8]	Zhida Liang	5A33 Al–AZ31B Mg	D 12 mm	60 MPa	1–20 s	180 MPa	>10 s	2200			

(Continued)

TABLE 6.1 (*Continued*)

Collective literatures of friction welding between ferrous and nonferrous metal alloys.

Ref	Author	Material	Specimen Dimension, Diameter (*D*), Length (*L*)	Friction Force/ Pressure	Friction Time	Upsetting force/Forge Pressure	Upsetting/ Forge time	Rotational Speed (rpm)	Burnoff Length	Braking Time	Interlayer
[9]	Lakshminarayanan	CP Ti–AZ31B Mg	D 15 mm, L 80 mm	20 MPa	4 s	50 MPa	8 s	1100			
[10]	Rajesh Jesudoss Hynes	AISI 1030– AA6063– 6SiC$_P$-3Gr$_P$	D 10 mm	3–5 bar	3–7 s			800–1600			
[11]	Della Rovere			4 MPa	24 s	4 MPa	4 s	200	7.5 mm	1 s	
	Prashanth	Ti-6Al-4V		275 MPa		550 MPa		150	3 mm		
[12]	Uday	AA6061 alloy– alumina– YSZ composite	D 16 mm	5 kN	30 s	30 MPa	40 s	630–2500			
[13]	Dey	Ti–304L SS	Ti-18 mm D 304L SS-14 mm D	100– 200 MPa		450 MPa	5 s	1500			

(Continued)

TABLE 6.1 (Continued)

Collective literatures of friction welding between ferrous and nonferrous metal alloys.

Ref	Author	Material	Specimen Dimension, Diameter (D), Length (L)	Friction Force/Pressure	Friction Time	Upsetting force/Forge Pressure	Upsetting/Forge time	Rotational Speed (rpm)	Burnoff Length	Braking Time	Interlayer
[14]	Mrityunjoy	High-nitrogen steel	D 15 mm, L 60 mm	20 kN		20–60 kN	10 s	1600	5 mm		
[15]	Zhida	Al1060–Mg AZ31B	D 12 mm, L 60 mm	30–120 MPa	10 s	60–150 MPa	10 s	2800	1.3–18.5 mm		
[16]	Ambroziak	W–Nb pseudoalloy	D 30 mm	23–71 N/mm²	3.5–25 s	82–203 N/mm²	8–100 s	1500		0.5 s	
[17]	Serdar Mercan	AISI2205–AISI1020	D 12 mm	30–50 MPa	4–8 s	60–100 MPa	2–4 s	1300–1500			
[18]	R. Kumar	Ti-6Al-4V–SS304L	D 20 mm	8–12 N/mm²	1–2 s	14–40 N/mm²	6–8 s	1500	7.6–11 mm		Oxygen-free copper
[19]	Meshram	Fe–Ti, Ti–Cu, Fe–Cu, Fe–Ni, Cu–Ni		3 kN		5 kN		1000	3–5 mm		
[20]	Antonio	Ti-6Al-4V (TiC particulate reinforced)	D 6 mm, L 30 mm	136, 272 MPa	–	–	–	1500–4500	–	–	–

(Continued)

TABLE 6.1 (*Continued*)
Collective literatures of friction welding between ferrous and nonferrous metal alloys.

Ref	Author	Material	Specimen Dimension, Diameter (*D*), Length (*L*)	Friction Force/ Pressure	Friction Time	Upsetting force/Forge Pressure	Upsetting/ Forge time	Rotational Speed (rpm)	Burnoff Length	Braking Time	Interlayer
[21]	Ananthapadmanaban	Mild steel– stainless steel	D 15 mm, L 75 mm	80, 160 MPa		160, 280 MPa		1500	1 mm, 2 mm		
[22]	Arivazhagan	AISI4140– AISI304		37.5 kN		50 kN		1500	5 mm, 12 mm		
[23]	Sare Celik	AISI4140– AISI1050	D 10 mm, L 80 mm	81, 121.5, 162, 202.5 MPa	4, 6, 8 s	162 MPa	14 s	3000			
[24]	Ahmet Hascalik	Al$_2$O$_3$- reinforced Al6160– SAE1020	D 12 mm, L 35 mm	10 MPa	8 s	5 MPa	4 s	2500			
											(*Continued*)
[25]	Jayabarath	Sintered P/M steel– wrought copper	D 20 mm, L 60 mm	47.74, 63.69, 95.5 MPa		63.69, 127.34, 95.5 MPa		1500	2, 3, 4 mms		
											(*Continued*)

TABLE 6.1 (Continued)
Collective literatures of friction welding between ferrous and nonferrous metal alloys.

Ref	Author	Material	Specimen Dimension, Diameter (D), Length (L)	Friction Force/Pressure	Friction Time	Upsetting force/Forge Pressure	Upsetting/Forge time	Rotational Speed (rpm)	Burnoff Length	Braking Time	Interlayer
[26]	Pengkang Zhao	TC11–TC17		50 MPa	3.6 s						
[27]	Udayakumar	UNS S32760	D 16mm, L 100mm	40, 80, 120MPa		122, 146, 170MPa		1000, 1500, 2000	2, 4, 6mm		
[28]	Srinivasan	AZ31B (Mg–Al$_2$O$_3$–Ca)	D 18mm, L 75mm	25, 35, 45MPa	3–6s	45, 55, 65MPa	5 s	1500			
[29]	Hazman Seli	Alumina–mild steel	D 10mm, L 50mm	29MPa	2–20s		<2s	900			Al 1100
[30]	S. R. Sundara Bharathi	AA2024–AA6061	D 12mm, L 75mm	0.43, 0.78, 1.24 ton	3–5s		3–5s	1100, 1200			
[31]	Sori Won	AA6063–AA2017	D 12mm,	1.2MPa		2.5MPa		2000	2.3 mm		
[32]	Furkan Sarsilmaz	AISI2205–Armor 500	D 12mm	60–80MPa	4–8s	120MPa	6s	1800			
[33]	Longwei Pan	AA1060–T2 Cu	Al D 25 mm, L 120mm Cu D 21mm, L 100mm	75MPa	3s	100MPa	5 s	1500			

Similar and Dissimilar Metal Joints by FW   141

Mild steel and Grad-2 Ti dissimilar joint were evaluated with optimized parameters by Prasanthi et al. They summarized that fine intermetallic phase was found in isolated regions at friction-welded interface and explored different phase formation based on time and temperature [14]. Selvamani et al. characterized the various mechanical properties of friction-welded AISI1035 steels and reported that the range of hardness is higher at heat-affected zone (HAZ) region compared with parent metal [15].

Nickel-free high-nitrogen steel friction-welded joints were examined by Mrityunjoy Hazra et al. [16]. The weld samples exhibited poor toughness with increase in upsetting pressure and tensile strength, and pitting corrosion resistance was lower than the base metal. Serdar Mercan et al. [17] made AISI2205–AISI1020 joints with friction pressure of 30 and 50 MPa) and forging pressure of 60 and 100 MPa with respect to 1300 and 1500 rpm. The highest fatigue strength was obtained which the joint was made with 1500 rpm and 30 MPa friction force. It has been identified that the tension and fatigue strength were reduced with increased friction time and friction force. Meshram et al. [18] dealt the effect of interaction time with the pure form of Fe–Ti, Ti–Cu, Fe–Cu, Fe–Ni, and Cu–Ni friction-welded joints. They increased interaction period and observed the formation of intermetallic layer eutectoid system in Fe–Ti and Cu–Ti and insoluble systems on Fe–Cu. A mild steel and stainless steel combination mechanical behavior was investigated by Ananthapadmanaban et al. They observed maximum hardness values around HAZ side for all samples. The maximum tensile strength of 629 MPa was found when the sample has 80 MPa friction force, 160 MPa upsetting pressure, and 1 mm burnoff length. During tensile testing, all the samples were broken in welded region, and a range of 2.4%–10% elongation was measured across the welded region with the help of an extensometer [19]. Arivazhagan et al. [20] studied corrosion behavior on AISI4140 and AISI304 dissimilar joint weldment at temperature of 500°C and 550°C in molten salt (Na_2SO_4 and V_2O_5) environment under cyclic condition. At the end of hot corrosion treatment, it was predicted that the formation of scale was high on AISI4140 side than AISI304 side. It has been finally concluded that to achieve better hot corrosion resistance, optimum burnoff length has to be maintained because it reduces intermetallic compounds and maintains uniform microstructure. AISI4140 and AISI1050 steel joints exhibited high hardness and grain size reduction at HAZ region due to increase of chromium ratio and martensitic structure. The maximum tensile strength was acquired in one of the specimens, which was 6% higher than actual AISI1050. The optimum welding parameters are predicted from this work, that is, 3000 rpm rotational speed, 81 MPa friction pressure, 6 s friction time, 162 MPa forge pressure, and 14 s upsetting time [21]. Udayakumar et al. [22] conducted an experimental investigation on UNS S32760 superduplex stainless steel friction-welded joint and developed three-level central composite design matrix based on response surface methodology which consists of four factors to predict optimum of FW process.

6.4   FRICTION WELDING BETWEEN NONFERROUS METAL ALLOYS

Yanni Wei et al. [23] made continuous drive friction-welded Al/Cu alloy joint with different friction time and characterized by using various modern techniques and stated that the interface temperature range of 648–723 K is suitable for different welding parameters. Friction weldability of 5A33 Al alloy to AZ31B Mg alloy was studied by Zhida Liang et al. [24]. In this work, the friction time varied and other parameters were kept constant for the experiment. It has been mentioned that reaction layer thickness was increased when the friction time increased, and the friction interface of all samples was failed during tensile test [25]. Also working with Al1060–AZ31B Mg combination, they observed that all samples were failed under tension test due to formation of intermetallic compounds. Sori Won et al. [26] have chosen different aluminum alloys such as AA6063 and AA2017 for FW and analyzed the characteristics of galvanic effect of biased corrosion behaviors at welded region. The open-circuit potential results showed improvement of corrosion behavior in AA6063. Lakshminarayanan et al. [27] observed microstructure of all regions of dissimilar friction-welded AZ31B Mg–Ti joint. The calculated joint efficiency is 104% between Mg–Ti joint. In addition to that friction-welded Mg–Ti joint, fracture was propagated from magnesium side during tensile test.

Uday et al. [28] studied microstructural consequence on plastic deformation among AA6061 alloy and alumina–YSZ composite. They concluded that the grain growth rate was improving while increasing the rotational speed parameter in FW. At low rotational speed, the rate of grain growth was low, minimizes the defect density, and increases grain size. Ambroziak et al. [29] produced niobium and tungsten pseudoalloy joints with optimized parameters. Use of copper as an interlayer between tungsten pseudoalloy D18 with niobium showed better microstructure and tensile strength. Physical and morphological characteristics of TiC particle–reinforced Ti-6Al-4V metal composite friction-welded joints were examined by Antonio et al. [30]. They identified smaller particle size and homogeneous distribution of TiC particles at transformed recrystallization zone than base metal under high rotational speed and pressure. Between TC11 and TC17 forged titanium alloy joint was prepared through linear FW. The tensile-tested specimens' fracture morphology showed similar stain rate of $10^{-3}s^{-1}$ at the center of joint. The center area of titanium alloy joint always failed at TC11 base metal side [31]. Srinivasan et al. [32] analyzed the possible parameters for effective FW in AZ31B Mg–Al_2O_3–Ca composite. The optimum parameters identified from the research were 35 MPa friction pressure, 5 s friction time, 55 MPa forge pressure, and 5 s forge time. Sundara Bharathi et al. [33] investigated on AA2024 and AA6061 similar and dissimilar friction-welded joint. They noted that yield and ultimate tensile strength of the joint is poorer than base metal alloy and found variations in hardness measurement on both similar and dissimilar joints. Longwei Pan et al. [34] used steel ring as an external tool during FW of AA1060-T2 Cu joint. The microstructure was captured on thermomechanically affected zone, heat-affected zone, and dynamic recrystallization zone of welded area. The tensile properties result decreased with increased intermetallic compound thickness of weldment.

6.5 FINITE ELEMENT MODEL IN FRICTION WELDING

Jedrasiak et al. [35] developed 2D thermal model linear FW and applied on Ti-6Al-4V instrumented welded joint. The elevated temperature region was gauged and found closer to the weld interface. Finally, they stated that developed thermal model was computationally proficient for applications. Mohammed Asif et al. [36] developed UNS S31803 duplex stainless steel 3D finite element (FE) model (FEM) in ANSYS software and analyzed the effect of temperature distribution and shortening during FW. They have concentrated on three levels of FW by considering the parameters such as heating pressure, heating time, upsetting pressure, and upsetting time. They compared the simulated results of peak temperature and axial shortening with actual experiment, and it showed similar results. They found that, due to high forging temperature, large amount of axial shortening was obtained in upsetting stage compared with heating stage. Friction-welded tubular CrMoV component simulation has been carried out by Bennet [37] with the help of DEFORM-2D (v10.2). The results indicated that the HAZ width was reduced at higher forge pressure [38]. FGH96 superalloy FEM was built in MSC Marc software by Longfei Nie et al. The simulations were focused on dynamic recrystallization (DRX) occurrences during inertia FW process. The simulated results were showed, and fraction of DRX decreases due to increasing distance of friction interface. Gianluca Buffa et al. [39] developed linear friction welding (LFW) machine to validate shear coefficient on aluminum alloy AA2011-T3 with an aid of DEFORM-3D™ software thermal and thermomechanical numerical FE model. They have used six case studies to determine the coefficient of friction and comparatively validated the test results between numerically calculated and experimentally measured records. Grant et al. [40] built FEM to predict temperature, residual stress, and strain rate for inertia friction-welded RR1000 nickel-based superalloy. The results of hoop stresses were very significant near the weldment, and it has been found that high welding pressure increases the strain rate, and strain rate region was eliminated into flash region by increasing the pressure.

6.6 CONCLUSION

From the above thorough literature survey, the following can be drawn:

1. The selection of welding parameters plays an important role during FW process because each parameters directly influence heat generation, structural integrity, and microstructure.
2. The better joint efficiency was achieved by increasing friction time and forge pressure in all metal alloys.
3. Introduction of intermediate layer in between the weld interfaces improves the joint efficiency and reduces the formation of intermetallic compounds.
4. It is observed from the authors' description that prediction and measurement of residual stress at welded region is quite difficult because it varies based on the process parameters. Therefore, further investigation on residual stress measurement and postweld heat treatment (PWHT) is needed.
5. Formation of intermetallic compounds and microstructure transformation of high-temperature ferrous alloys are yet to be explored.

REFERENCES

1. W. Li, A. Vairis, M. Preuss, and T. Ma, "Linear and rotary friction welding review," *Int. Mater. Rev.*, vol. 6608, no. 2, pp. 71–100, February, 2016.
2. M. Kimura, K. Suzuki, M. Kusaka, and K. Kaizu, "Effect of friction welding condition on joining phenomena and mechanical properties of friction welded joint between 6063 aluminium alloy and AISI 304 stainless steel," *J. Manuf. Process.*, vol. 26, pp. 178–187, 2017.
3. R. Winiczenko and M. Kaczorowski, "Friction welding of ductile cast iron using interlayers," *Mater. Des.*, vol. 34, pp. 444–451, 2012.
4. H. C. Dey, M. Ashfaq, A. K. Bhaduri, and K. P. Rao, "Joining of titanium to 304L stainless steel by friction welding," *J. Mater. Process. Technol.*, vol. 209, no. 18–19, pp. 5862–5870, 2009.
5. S. D. Meshram and G. M. Reddy, "Friction welding of AA6061 to AISI 4340 using silver interlayer," *Def. Technol.*, vol. 11, pp. 292–298, 2015.
6. R. Kumar and M. Balasubramanian, "Experimental investigation of Ti-6Al-4V titanium alloy and 304L stainless steel friction welded with copper interlayer," *Def. Technol.*, vol. 11, pp. 65–75, 2015.
7. A. Hascalik, "Effect of particle size on the friction welding of Al2O3 reinforced 6160 Al alloy composite and SAE 1020 steel," *Mater. Des.*, vol. 28, pp. 313–317, 2007.
8. K. Jayabharath, M. Ashfaq, P. Venugopal, and D. R. G. Achar, "Investigations on the continuous drive friction welding of sintered powder metallurgical (P/M) steel and wrought copper parts," *Mater. Sic. Eng.*, vol. 455, pp. 114–123, 2007.
9. G. Vairamani, T. S. Kumar, S. Malarvizhi, and V. Balasubramanian, "Application of response surface methodology to maximize tensile strength and minimize interface hardness of friction welded dissimilar joints of austenitic stainless steel and copper alloy," *Trans. Nonferrous Met. Soc. China*, vol. 23, no. 8, pp. 2250–2259, 2013.
10. H. Seli, M. Zaky, A. Izani, E. Rachman, and Z. Arifin, "Characterization and thermal modelling of friction welded alumina – mild steel with the use of Al 1100 interlayer," *J. Alloys Compd.*, vol. 506, no. 2, pp. 703–709, 2010.
11. Akram, J., Kalvala, P. R., Misra, M., & Charit, I. (2017). Creep behavior of dissimilar metal weld joints between P91 and AISI 304. *Materials Science and Engineering*: A, 688, 396–406.
12. F. Sarsilmaz, I. Kirik, and S. Batı, "Microstructure and mechanical properties of armor 500/AISI2205 steel joint by friction welding," *J. Manuf. Process.*, vol. 28, pp. 131–136, 2017.
13. J. A. James and R. Sudhish, "Study on effect of interlayer in friction welding for dissimilar steels : SS 304 and AISI 1040," *Procedia Technol.*, vol. 25, pp. 1191–1198, 2016.
14. T. N. Prasanthi, C. Sudha, S. Saroja, N. N. Kumar, and G. D. Janakiram, "Friction welding of mild steel and titanium : optimization of process parameters and evolution of interface microstructure," *JMADE*, vol. 88, pp. 58–68, 2015.
15. S. T. Selvamani, K. Palanikumar, K. Umanath, and D. Jayaperumal, "Analysis of friction welding parameters on the mechanical metallurgical and chemical properties of AISI 1035 steel joints," *Mater. Des.*, vol. 65, pp. 652–661, 2015.
16. M. Hazra, K. S. Rao, and G. M. Reddy, "Friction welding of a nickel free high nitrogen steel: Influence of forge force on microstructure, mechanical properties and pitting corrosion resistance," *J. Mater. Res. Technol.*, vol. 3, no. 1, pp. 90–100, 2014.
17. S. Mercan, S. Aydin, and N. Özdemir, "Effect of welding parameters on the fatigue properties of dissimilar AISI 2205- AISI 1020 joined by friction welding," *Int. J. Fatigue*, vol. 81, no. 1, pp. 78–90, 2015.

18. S. D. Meshram, T. Mohandas, and G. M. Reddy, "Friction welding of dissimilar pure metals," *J. Mater. Process. Technol.*, vol. 184, no. 1–3, pp. 330–337, 2007.
19. D. Ananthapadmanaban, V. S. Rao, N. Abraham, and K. P. Rao, "A study of mechanical properties of friction welded mild steel to stainless steel joints," *Mater. Des.*, vol. 30, no. 7, pp. 2642–2646, 2009.
20. N. Arivazhagan, S. Singh, S. Prakash, and G. M. Reddy, "High temperature corrosion studies on friction-welded dissimilar metals," *Mater. Sci. Eng.*, vol. 132, pp. 222–227, 2006.
21. S. Celik and I. Ersozlu, "Investigation of the mechanical properties and microstructure of friction welded joints between AISI 4140 and AISI 1050 steels," *Mater. Des.*, vol. 30, no. 4, pp. 970–976, 2009.
22. T. Udayakumar, K. Raja, A. T. Abhijit, and P. Sathiya, "Experimental investigation on mechanical and metallurgical properties of super duplex stainless steel joints using friction welding process," *J. Manuf. Process.*, vol. 15, pp. 558–571, 2013.
23. Y. Wei, J. Li, J. Xiong, and F. Zhang, "Investigation of interdiffusion and intermetallic compounds in Al – Cu joint produced by continuous drive friction welding," *Eng. Sci. Technol. an Int. J.*, vol. 19, no. 1, pp. 90–95, 2016.
24. Z. Liang, G. Qin, P. Geng, F. Yang, and X. Meng, "Continuous drive friction welding of 5A33 Al alloy to AZ31B Mg alloy," *J. Manuf. Process.*, vol. 25, pp. 153–162, 2017.
25. Liang, Z., Qin, G., Wang, L., Meng, X., & Li, F. (2015). Microstructural characterization and mechanical properties of dissimilar friction welding of 1060 aluminum to AZ31B magnesium alloy. *Materials Science and Engineering*: A, 645, 170–180.
26. S. Won, B. Seo, J. M. Park, H. K. Kim, K. H. Song, and S. Min, "Corrosion behaviors of friction welded dissimilar aluminum alloys," *Mater. Charact.*, vol. 144, no. June, pp. 652–660, 2018.
27. A. K. Lakshminarayanan and R. Saranarayanan, "Characteristics of friction welded AZ31B magnesium – commercial pure titanium dissimilar joints," *J. Magnes. Alloy.*, vol. 3, pp. 315–321, 2015.
28. M. B. Uday, M. N. Ahmad-Fauzi, A. M. Noor, and S. Rajoo, "An insight into microstructural evolution during plastic deformation in AA6061 alloy after friction welding with alumina-YSZ composite," *Mech. Mater.*, vol. 91, no. P1, pp. 50–63, 2015.
29. A. Ambroziak, M. Korzeniowski, P. Kustro, and M. Winnicki, "Friction welding of niobium and tungsten pseudoalloy joints," *Int. J. Refract. Metals Hard Mater.*, vol. 29, pp. 499–504, 2011.
30. A. A. M. da Silva, A. Meyer, J.F. Dos Santos, C. E. F. Kwietniewski, and T. R. Strohaecker, "Mechanical and metallurgical properties of friction-welded TiC particulate reinforced Ti – 6Al – 4V," *Compos. Sci. Technol.*, vol. 64, pp. 1495–1501, 2004.
31. P. Zhao and L. Fu, "Strain hardening behavior of linear friction welded joints between TC11 and TC17 dissimilar titanium alloys," *Mater. Sci. Eng. A*, vol. 621, pp. 149–156, 2015.
32. M. Srinivasan, C. Loganathan, V. Balasubramanian, Q. B. Nguyen, M. Gupta, and R. Narayanasamy, "Feasibility of joining AZ31B magnesium metal matrix composite by friction welding," *Mater. Des.*, vol. 32, no. 3, pp. 1672–1676, 2011.
33. S. R. S. Bharathi, R. Rajeshkumar, A. R. Rose, and V. Balasubramanian, "Mechanical properties and microstructural characteristics of friction welded dissimilar joints of aluminium alloys," *Mater. Today Proc.*, vol. 5, no. 2, pp. 6755–6763, 2018.
34. Pan, L., Li, P., Hao, X., Zhou, J., & Dong, H. (2018). Inhomogeneity of microstructure and mechanical properties in radial direction of aluminum/copper friction welded joints. *Journal of Materials Processing Technology*, 255, 308–318.
35. P. Jedrasiak, H. R. Shercliff, A. R. Mcandrew, and P. A. Colegrove, "Thermal modelling of linear friction welding," *Mater. Des.*, vol. 156, pp. 362–369, 2018.

36. M. A. M, K. A. Shrikrishana, and P. Sathiya, "Finite element modelling and characterization of friction welding on UNS S31803 duplex stainless steel joints," *Eng. Sci. Technol. an Int. J.*, vol. 18, pp. 704–712, 2015.

37. C. Bennett, "Finite element modelling of the inertia friction welding of a CrMoV alloy steel including the effects of solid-state phase transformations," *J. Manuf. Process.*, vol. 18, pp. 84–91, 2015.

38. L. Nie, L. Zhang, Z. Zhu, and W. Xu, "Microstructure evolution modeling of FGH96 superalloy during inertia friction welding process," *Finite Elem. Anal. Des.*, vol. 80, pp. 63–68, 2014.

39. G. Buffa, M. Cammalleri, D. Campanella, and L. Fratini, "Shear coefficient determination in linear friction welding of aluminum alloys," *Mater. Des.*, vol. 82, pp. 238–246, 2015.

40. B. Grant, M. Preuss, P. J. Withers, G. Baxter, and M. Rowlson, "Finite element process modelling of inertia friction welding advanced nickel-based superalloy," *Mater. Sci. Eng. A*, vol. 514, pp. 366–375, 2009.

7 3D Bioprinting in Pharmaceuticals, Medicine, and Tissue Engineering Applications

Atul Babbar
Shree Guru Gobind Singh Tricentenary University,
Gurugram, Haryana, India

Vivek Jain, and Dheeraj Gupta
Thapar Institute of Engineering and Technology

Chander Prakash and Sunpreet Singh
Lovely Professional University

Ankit Sharma
Chitkara College of Applied Engineering
Chitkara Univeristy

CONTENTS

7.1 Introduction 148
7.2 3D Printing in Pharmaceuticals 150
7.3 3D Printing in Medicine 152
    7.3.1 Materials 152
    7.3.2 In situ 3D Bioprinting 153
        7.3.2.1 Biomimicry 154
        7.3.2.2 Independent Self-Assembly 154
        7.3.2.3 Miniature Tissue Blocks 155
    7.3.3 Bioscaffolding 155
7.4 Conclusion 157
References 157

7.1 INTRODUCTION

The concept of the three-dimensional (3D) printing (3DP) has evolved in the early 1970s, when Pierre A. L. Ciraud suggested using high energy beam for the solidification of the layer, primarily the application of the powdered material. 3DP is a technique in which a 3D object is fabricated by either depositing or fusing plastic, powder, ceramics, metals, and biomedical devices such as implants, prosthetics, stents, or even biological living tissue cells and is known as additive manufacturing (AM) or rapid prototyping technique [1]. 3DP has enabled us to a high level of customization as per the requirement in various biomedical fields such as pharmaceutical, medicine, artificial organs, bioscaffolding, and so on [2]. Figure 7.1 shows the schematic representation of the four fundamental categories of the 3DP. Fused deposition modeling (FDM) is a layer-by-layer material deposition process of fabricating

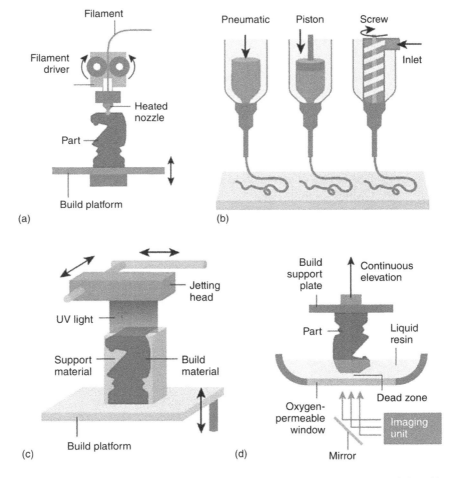

FIGURE 7.1 Common 3DP methods. (a) Fused deposition modeling, (b) direct ink writing, (c) inkjet, and (d) projection stereolithography [4]. 3DP, three-dimensional printing.

3D Bioprinting

a 3D object with the help of extruder [3]. The filament material is melted in the nozzle, and soften material is forced to extrude out from the nozzle end on the build tray as shown in Figure 7.1a. In this way, 3D structures are fabricated using different materials such as acrylonitrile butadiene styrene (ABS) and polylactic acid (PLA). Direct ink writing (DIW) works on the same principle as of FDM.

The material used in DIW is viscoelastic ink instead of thermoplastic material used in FDM [5]. Ink is extruded out from the nozzle under high pressure, and hence, 3D object is fabricated (refer to Figure 7.1b. Inkjet technology spreads tiny drops of the ink which has low viscosity. The 3D structure is fabricated on the build platform with the help of UV light and ink nozzle as shown in Figure 7.1c. Furthermore, multiple printing heads can be used to print using different kinds of materials and are referred to as polyjet [6]. The schematic illustration of the projection stereolithography (pSLA) is shown in Figure 7.1d. SLA is a point source illumination method in which a rastering laser is used to polymerize a liquid polymer, subsequently cross-linking the polymer and hence fabricating the 3D object [7].

DIW method has an advantage over the other methods as it does not require nozzles and hence eliminates the problem of clogging. In pSLA, a patterned ultraviolet light is projected over the surface of a polymer (photosynthesized resin). The build support plate is continuously moved upward. The complex 3D structures can be easily fabricated using this process. The biomedical applications of 3DP have been shown in Table 7.1.

TABLE 7.1
Biomedical Applications of the 3DP, Products, and Techniques Available [8–12]

Applications	Products Available	Techniques	3D Printing Companies
Dentistry	Surgical guides, dental implants, and restorations such as dentures, prosthesis, and braces	SLA, SLS, binder jet printing	Stratasys, Envision TEC, 3D Systems, FormLabs
Tissues, organs, and models	Tissue analogous for implantation, disease models, and drug testing	FDM, SLS, binder jet printing	Organovo, Envision TEC, Biobots, n3D Biosciences
Anatomical models	Surgical planning tools for training and education	SLS, FDM, SLA, binder jet printing	3D Systems, Stratasys, Materialise
Pharmaceuticals	Controlled drug delivery, oral disintegrating tablets, personalized medicine	SLS, FDM, binder jet printing	SPRITAM of Aprecia Pharmaceuticals
Medical devices	Surgical instruments, implants, prosthesis, orthoses, hearing aids	SLS, FDM, SLA, binder jet printing, SLM	Stratasys, Envision TEC, 3D Systems

Notes: FDM, fused deposition modeling; SLA, stereolithography; SLM, selective laser melting; SLS, selective laser sintering.

7.2 3D PRINTING IN PHARMACEUTICALS

The invention of the 3DP has emerged as a paradigm shift in the design and manufacturing of the tablets. Now, new tailor-made tablets can be fabricated with different shape, size, and characteristics. 3DP has completely revolutionized the process of drug development and delivery system for human beings. It has been revealed that torus-shape tablets have higher acceptance score in comparison with the other shape tablets as shown in Figure 7.2.

3DP technology has enabled us to manufacture the tablets with short life and lower stability with on-demand printing and can be used on primary, secondary, and tertiary health centers. The on-demand manufacturing capability could lead to the improved discharge time and easy availability of the medicines. Not only hospitals but also 3DP can be used for space, military, and third world operations. The flexibility in 3DP allows to quickly modify or change the dosages as per the results achieved during in vivo and in vitro clinical trials. Infill percentage (i.e., from 0% to 100%) of the tablets can be easily controlled along with the dimensions of the tablet [14,15]. 3DP systems are compact and can be easily integrated into the laboratory. Hence, they can be used for the data collection and testing of the tablets with different iterations [16,17].

Binder jet technology (BJT) is another important 3DP process used in the pharmaceutical industry. In this technology, the powder material is agglomerated in the form of a layer on a building platform [18–21]. The schematic illustration of the BJT process has been shown in Figure 7.3. Powder delivery platform is also raised which is located to the adjacent of the fabrication platform. The printer nozzle moves in the x-y direction, and fabrication platform moves in the z-direction. The roller moves the powder forward to the fabrication platform. The liquid is sprayed and binds the powder from a 3D structure. The first tablet printed using BJT is Spritam® and has been proved to be successful in the pharmaceutical industry. It provides multiple advantages from controlled dissolution to the fast dissolution characteristics. Table 7.2 shows the literature review of the BJT for various types of compound formulations.

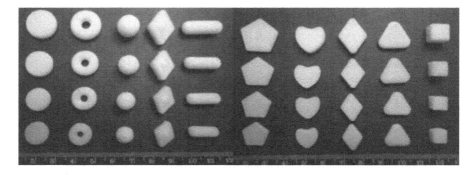

FIGURE 7.2 Range of shapes and sizes of 3D printing [13].

3D Bioprinting 151

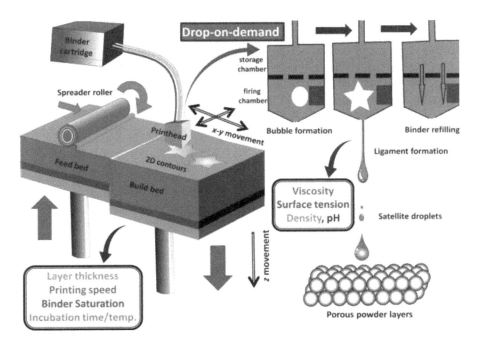

FIGURE 7.3 Schematic representation of powder jet technology 3D printing [22].

TABLE 7.2
Selected Literature Review of Binder Jet Technology for Different Formulations

Type of Formulation	Drug Compound	Binder/Solvent	Drug Added	Reference
Predetermined drug release tabular device	Methylthioninium chloride Alizarin yellow	Chloroform PCL	Ink	[23]
Near-zero-order release pattern (cubic dosage)	Pseudoephedrine	PVP K 17 PVP K17+TEC	Ink	[24]
Fast disintegrating tablets	Paracetamol	PVP K30 Ethanol Water	Powder bed	[25]
Delayed release enteric-coated tablets (Depakote Sprinkles)	Paracetamol	Ethylcellulose+ethanol	Powder bed	[26]
Delayed response tablets activated by erosion mechanism	Chlorphenamine Fluorescein	Ethanol Acetone PVP+water	Ink	[27]
Zero-order release tablets	Paracetamol	Ethylcellulose	Powder bed	[28]

PCL, polycaprolactone; PVP, polyvinylpyrrolidone.

7.3 3D PRINTING IN MEDICINE

3DP has now completely revolutionized the field of medicine with state-of-the-art technologies, greater flexibility in the product design, and its manufacturing. The basic 3DP can be divided into three components, namely (i) hardware (printer), (ii) software (conversion of CAD images into a format that a printer can recognize), and (iii) material [29]. Different types of biomaterials, bioprinting, and bioscaffolding have been used for the requisite applications, which are discussed in the next subsections.

7.3.1 MATERIALS

The nonviable materials that are used to make an interface with the biological tissues are known as biomaterials. These materials have numerous applications ranging from drug delivery, implants, and prosthesis to tissue engineering (TE) applications [30–32]. By 2020, the biomaterials market is expected to overcome US$130 billion with a 16% yearly growth rate. Table 7.3 shows some common biomaterials used in clinical aspects.

Several metals and their alloys such as titanium, stainless steel, and cobalt-based alloys have been utilized in the medical industry owing to their excellent biocompatibility and mechanical characteristics [37–40]. For a long duration, titanium is generally preferred for the biomedical implants. Titanium has the ability to passivate itself and display of biocompatibility and osseointegration owing to the formation of a thin oxide layer [41]. Figure 7.4a and b shows the 3D printed titanium-based scaffolds (Ti-24Nb-4Zr-8Sn) and biodegradable magnesium stents.

TABLE 7.3
Some Common Biomaterials with Their Applications [12,33–36]

Biomaterial	Applications
Titanium alloys	Heart valves, dental implants, spinal cages, etc.
Polyester textile	Heart valves and vascular grafts
Cobalt–chromium alloy	Pacemaker leads, stents, and heart valves
Expanded PTFE	Heart valves and vascular grafts
Polyurethane	Pacemaker insulation
NiTiNOL	Shape memory applications
Silicones	Ophthalmological devices and soft tissues augmentation
Stainless steel	Stents and implants
PMMA	Intraocular lenses, bone cement
PEEK	Spinal cages
Calcium phosphate	Bioactive surfaces and substitutes of bone
Carbon	Heart valves

NiTiNOL, nickel–titanium Naval Ordnance Laboratory; PEEK, polyether ether ketone; PMMA, polymethyl methacrylate; PTFE, polytetrafluoroethylene.

3D Bioprinting

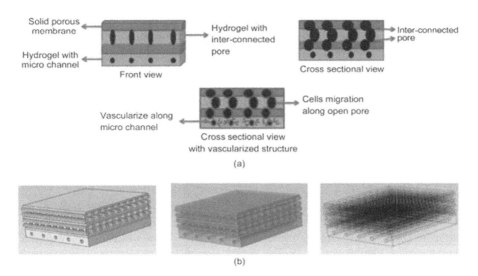

FIGURE 7.4 Scaffolds fabricated using additive manufacturing process (under a Creative Commons license) [42]. (a) Front and cross-sectional views of micro channel. (b) Complete view of fabricated scaffolds for vascularization.

Implants for load-bearing applications have been fabricated using tantalum and cobalt–chromium (CoCr) alloys. The results showed by tantalum-based porous structures are superior to the CoCr-based structures [43,44]. Stainless steel is also used for the load-bearing applications. Moreover, it is easily available, low cost, and corrosion resistance nature. Bioceramics have tremendous applications in the body joints repair. Similarly, alumina can also be used owing to high hardness and low frictional coefficient. Zirconia offers lower wear rate during its life span. Silica-based bioglasses have wider applications in the biomedical industry. In 1969, Larry Hench introduced bioactive glasses that can be used as a biomaterial in the human body [45–47]. It has been observed that glass forms a hydroxyapatite (HAP) layer on its surface when exposed to the biological fluid, and this allows to interact with the body cells. In the medical industry, biodegradable polymer materials such as PLA, polyether ether ketone (PEEK), polycaprolactone, photocurable polymers, and gelatin-based polymers have been used [48–50]. Nowadays, bioinks (hydrogels) have been used for the 3D bioprinting, which has been discussed in Section 3.3.

7.3.2 In situ 3D Bioprinting

Different types of 3D bioprinting techniques have been used, which include cell-encapsulating hydrogel beads, photolithographic process, cavitation bubble generation, and print polymers, which use valve-based, laser, acoustic, and inkjet mechanisms, respectively, to deposit cells and hydrogels. In the valve-based method actuation frequency and duration of opening are important aspects of 3D bio printing. The laser-based mechanism works on the difference between the refractive indexes of media of cell and living cells. In acoustic-based technology, acoustic waves are used which has no effect on the health of tissues and allows a wider range of droplets [43]. In inkjet technology, cell aggregates are placed in the extruder in which complete process of printing is precisely controlled with computer [51,52]. The schematic representation of the 3DP approaches is shown in Figure 7.5.

3D bioprinting includes three approaches:

1. Biomimicry
2. Independent self-assembly (ISA)
3. Miniature tissue blocks

7.3.2.1 Biomimicry

The issues related to the material research, nanotechnology, and protocols of cell culture have been addressed using biomimicry. In 3D bioprinting, exact replicates of the native organs have been fabricated with the reproduction of cellular functionality. The microscale reproduction is an essential requirement for using this approach. All necessary information related to the microenvironment is vital to understand, which includes spatial arrangement, biochemical factors, aggregation of cells, biological forces, and so on.

7.3.2.2 Independent Self-Assembly

This approach uses a guide comprised of embryonic organ development process to drive the tissues produced using bioprinting. The information about organ and tissue genesis along with embryonic histogenesis is must so as to control the microenvironment required for bioprinting the tissues. ISA depends upon the cell's localization, functional, and structural characteristics. Furthermore, histogenesis and composition also play a

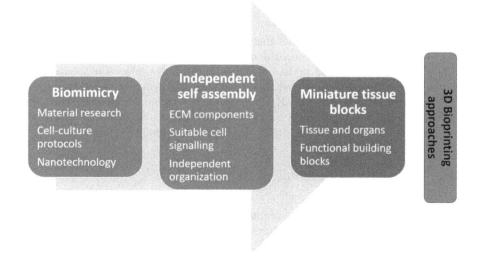

FIGURE 7.5 3D printing approaches. ECM, extracellular matrix.

3D Bioprinting

155

significant role. The cellular components produce their own independent pattern and signals to cause fusion to mimic the development process in the microenvironment.

7.3.2.3 Miniature Tissue Blocks

The group of cells combines or aggregates together to form a miniature tissue. The tissues that comprise of multiple building blocks are termed as miniature tissues. As we know, tissue itself is a smaller component, but it has been observed that miniature tissue is the basic structural component of the tissue. It can be easily understood with the example of the kidney nephron. These miniature tissues can be simultaneously fabricated to form a large tissue. Different characteristics of bioprinting such as material, design, imaging, and selection of cells play a crucial role in the final 3D biological model [53–61]. Bioinspired design approach can be used to fabricate the tissue in which miniature tissues organize itself and form a macrotissue. One another approach is replicating the exact tissue block followed by self-assembly which leads to the formation of the macrotissue. One of the important outcomes of such 3DP approaches is "organs-on-chips" model that is used in the in vitro screening of the various drugs and vaccines.

The schematic representation of the ex vivo cells printed using the 3D bioprinting technique is shown in Figure 7.6a. In situ bioprinting illustration is shown in Figure 7.6b on animal tissues (refer to Figure 7.6b). The future vision of in vivo direct tissue repair has been schematically represented in Figure 7.6c. The in situ 3DP is at the initial stage, but continuous progress in the technology and development of novel biomaterials could make it possible in future. Different approaches can be used while dealing with such processes [2,62–69]. Research has also been carried out on laser-assisted bioprinting in which bioink was used for printing a substitute for the bone. It has been noticed that tissue regeneration is significantly affected by cellular arrangements [68].

7.3.3 Bioscaffolding

In TE, scaffolds refer to the structures that have interconnected pores with the controlled pore size which can support the in vivo formation of the cells in living beings [70]. It has been observed that human beings have very less tendency to autoregenerate damaged organs. Here, 3D bioprinting plays a significant contribution toward the repair of the damaged organ. Figure 7.7 shows the 3D bioprinted tissue product that initially starts from the imaging technique (CT or MRI), and then a 3D CAD model is generated followed by the text-based command list, then the design is fed into the 3DP machine, and finally, a 3D structure is fabricated.

The scaffolds printed using 3DP technology have important functioning in the cell–cell and cell–extracellular matrix (ECM) interactions by providing a microenvironment (in vivo) for the cellular interaction. The scaffold should be designed in an optimal way to ensure a high level of cells proliferation, attachment, differentiation, and migration with biomimicry of the native ECM. Different factors should be considered while fabricating scaffolds which are biocompatibility, pore size and geometry, biomimicking, biodegradability, surface characteristics, and mechanical properties [72–74].

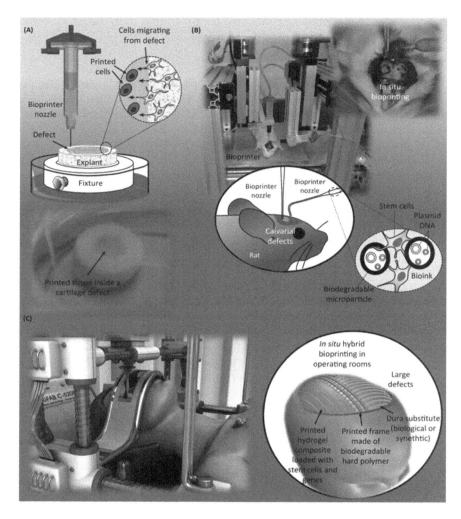

FIGURE 7.6 Direct repair of in situ 3D printing tissue repair [69].

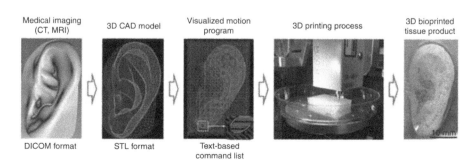

FIGURE 7.7 Process steps for 3D bioprinted tissue scaffold [71].

3D Bioprinting

"Bioink" term refers to the hydrogels (biomaterial) which allows to include cells within the AM process and printing a 3D constructs for the tissue regeneration in a controlled manner in a biological environment [75,76]. For easy extrusion of the bioink, the viscosity of the ink should be optimized to eliminate any chance of nozzle clogging and strand collapse. Bioinks require superior characteristics for stabilization of the scaffolds. The temperature selected for the fabrication of 3D structures should be taken considering the physiological behavior of the cells. Furthermore, the extrusion pressure needs to be selected considering the stability of the constructs. Strategies such as UV treatment and cross-linking with the chemical reaction often lead to the damage of included cells. To counter these concerns, different methods have been implemented by past researchers. Markstedt et al. [77] increased the viscosity of bioink, and 70% cell viability was observed having similar cartilage-like structure. The most popular basis biomaterial for bioink is alginate. However, to ensure suitable viscosity, other components have also been added. To support the alginate cross-linking, methylcellulose is used as a stabilizing material to fabricate 3D structures. Schultz et al. implemented this approach for the fabrication of 3D scaffolds using human mesenchymal stem cell (hMSC) encapsulated in cellulose bioink. Furthermore, the composite scaffolds have also been designed for TE applications. Luo et al. [78] used alginate/HAP (pH 9.5)-based core–shell scaffolds with mineralized, pure, and mixed alginate.

7.4 CONCLUSION

3DP has potential applications in the field of pharmaceuticals, medicine, and TE. It has been postulated that stem cells can be successfully harnessed at the birth stage and later can be used for the regeneration of the damaged tissues. Vast varieties of the biomaterial are under investigation using 3DP for manufacturing of implants, prosthesis, biomedical instruments, artificial organs, scaffolds, medicinal tablets, stents, and so on. Novel biodegradable materials are being developed to overcome the low life span concern of the metallic implants. Recently, tracheal splints have been successfully fabricated using 3DP to regulate normal breathing for tracheobronchomalacia patients. Different 3DP approaches and technologies have been practiced, but still, biological improvements in terms of cell viability, perfusability, and functionality are required. The concerns related to the size of constructs, printing speed, range of biomaterial, appropriate physical and mechanical properties, and nozzle clogging should also be addressed. No doubt, 3DP has completely revolutionized the medical industry with state-of-the-art printing of 3D structures. However, the current developments are primitive and need further innovation for innumerable opportunities for bioprinting and in vivo testing for vascular network and complete replacement of the damaged organs for on-demand requirements.

REFERENCES

1. C. Prakash, S. Singh, B.S. Pabla and M.S. Uddin, Synthesis, characterization, corrosion and bioactivity investigation of nano-HA coating deposited on biodegradable Mg-Zn-Mn alloy, *Surf. Coatings Technol.* 346 (2018), pp. 9–18.

2. D. Singh, A. Babbar, V. Jain, D. Gupta, S. Saxena and V. Dwibedi, Synthesis, characterization, and bioactivity investigation of biomimetic biodegradable PLA scaffold fabricated by fused filament fabrication process, *J. Brazilian Soc. Mech. Sci. Eng.* 41 (2019), pp. 121.

3. S. Singh, C. Prakash and S. Ramakrishna, 3D printing of polyether-ether-ketone for biomedical applications, *Eur. Polym. J.* 114 (2019), pp. 234–248.

4. R.L. Truby and J.A. Lewis, Printing soft matter in three dimensions, Nature 540 (2016), pp. 371–378.

5. J.A. Lewis, Direct ink writing of 3D functional materials, *Adv. Funct. Mater.* 16 (2006), pp. 2193–2204.

6. T. Boland, T. Xu, B. Damon and X. Cui, Application of inkjet printing to tissue engineering, *Biotechnol. J.* 1 (2006), pp. 910–917.

7. R. Gauvin, Y.-C. Chen, J.W. Lee, P. Soman, P. Zorlutuna, J.W. Nichol et al., Microfabrication of complex porous tissue engineering scaffolds using 3D projection stereolithography, *Biomaterials* 33 (2012), pp. 3824–3834.

8. S.V. Murphy and A. Atala, 3D bioprinting of tissues and organs, *Nat. Biotechnol.* 32 (2014), pp. 773–785.

9. X. Wang, M. Jiang, Z. Zhou, J. Gou and D. Hui, 3D printing of polymer matrix composites: A review and prospective, *Compos. Part B Eng.* 110 (2017), pp. 442–458.

10. H.N. Chia and B.M. Wu, Recent advances in 3D printing of biomaterials, *J. Biol. Eng.* 9 (2015), pp. 4.

11. B.C. Gross, J.L. Erkal, S.Y. Lockwood, C. Chen and D.M. Spence, Evaluation of 3D printing and its potential impact on biotechnology and the chemical sciences, *Anal. Chem.* 86 (2014), pp. 3240–3253.

12. A. Babbar, V. Jain and D. Gupta, Thermogenesis mitigation using ultrasonic actuation during bone grinding: A hybrid approach using CEM43°C and Arrhenius model, *J. Brazilian Soc. Mech. Sci. Eng.* 41 (2019), pp. 401.

13. A. Goyanes, M. Scarpa, M. Kamlow, S. Gaisford, A.W. Basit and M. Orlu, Patient acceptability of 3D printed medicines, *Int. J. Pharm.* 530 (2017), pp. 71–78.

14. A. Goyanes, P. Robles Martinez, A. Buanz, A.W. Basit and S. Gaisford, Effect of geometry on drug release from 3D printed tablets, *Int. J. Pharm.* 494 (2015), pp. 657–663.

15. J. Wang, A. Goyanes, S. Gaisford and A.W. Basit, Stereolithographic (SLA) 3D printing of oral modified-release dosage forms, *Int. J. Pharm.* 503 (2016), pp. 207–212.

16. S.A. Khaled, J.C. Burley, M.R. Alexander, J. Yang and C.J. Roberts, 3D printing of tablets containing multiple drugs with defined release profiles, *Int. J. Pharm.* 494 (2015), pp. 643–650.

17. M.A. Alhnan, T.C. Okwuosa, M. Sadia, K.-W. Wan, W. Ahmed and B. Arafat, Emergence of 3D printed dosage forms: Opportunities and challenges, *Pharm. Res.* 33 (2016), pp. 1817–1832.

18. E. Sachs, M. Cima, J. Cornie, D. Brancazio, J. Bredt, A. Curodeau et al., Three-dimensional printing: The physics and implications of additive manufacturing, *CIRP Ann. – Manuf. Technol.* 42 (1993), pp. 257–260.

19. M. Vaezi and C.K. Chua, Effects of layer thickness and binder saturation level parameters on 3D printing process, *Int. J. Adv. Manuf. Technol.* 53 (2011), pp. 275–284.

20. P. Nandwana, A.M. Elliott, D. Siddel, A. Merriman, W.H. Peter and S.S. Babu, Powder bed binder jet 3D printing of Inconel 718: Densification, microstructural evolution and challenges☆, *Curr. Opin. Solid State Mater. Sci.* 21 (2017), pp. 207–218.

21. V. Mironov, T. Boland, T. Trusk, G. Forgacs and R.R. Markwald, Organ printing: Computer-aided jet-based 3D tissue engineering, *Trends Biotechnol.* 21 (2003), pp. 157–161.

22. S. Barui, S. Mandal and B. Basu, Thermal inkjet 3D powder printing of metals and alloys: Current status and challenges, *Curr. Opin. Biomed. Eng.* 2 (2017), pp. 116–123.

23. B.M. Wu, S.W. Borland, R.A. Giordano, L.G. Cima, E.M. Sachs and M.J. Cima, Solid free-form fabrication of drug delivery devices, *J. Control. Release* 40 (1996), pp. 77–87.
24. C.-C. Wang, M.R. Tejwani (Motwani), W.J. Roach, J.L. Kay, J. Yoo, H.L. Surprenant et al., Development of near zero-order release dosage forms using three-dimensional printing (3-DP™) technology, *Drug Dev. Ind. Pharm.* 32 (2006), pp. 367–376.
25. D.-G. Yu, C. Branford-White, Y.-C. Yang, L.-M. Zhu, E.W. Welbeck and X.-L. Yang, A novel fast disintegrating tablet fabricated by three-dimensional printing, *Drug Dev. Ind. Pharm.* 35 (2009), pp. 1530–1536.
26. D.-G. Yu, C. Branford-White, Z.-H. Ma, L.-M. Zhu, X.-Y. Li and X.-L. Yang, Novel drug delivery devices for providing linear release profiles fabricated by 3DP, *Int. J. Pharm.* 370 (2009), pp. 160–166.
27. W.E. Katstra, R.D. Palazzolo, C.W. Rowe, B. Giritlioglu, P. Teung and M.J. Cima, Oral dosage forms fabricated by Three Dimensional Printing™, *J. Control. Release* 66 (2000), pp. 1–9.
28. D.G. Yu, X.L. Yang, W.D. Huang, J. Liu, Y.G. Wang and H. Xu, Tablets with material gradients fabricated by three-dimensional printing, *J. Pharm. Sci.* 96 (2007), pp. 2446–2456.
29. T.D. Ngo, A. Kashani, G. Imbalzano, K.T.Q. Nguyen and D. Hui, Additive manufacturing (3D printing): A review of materials, methods, applications and challenges, *Compos. Part B Eng.* 143 (2018), pp. 172–196.
30. A. Bandyopadhyay, S. Bose and S. Das, 3D printing of biomaterials, *MRS Bull.* 40 (2015), pp. 108–115.
31. U. Jammalamadaka and K. Tappa, Recent advances in biomaterials for 3D printing and tissue engineering, *J. Funct. Biomater.* 9 (2018), pp. 22.
32. W. Zhu, X. Ma, M. Gou, D. Mei, K. Zhang and S. Chen, 3D printing of functional biomaterials for tissue engineering, *Curr. Opin. Biotechnol.* 40 (2016), pp. 103–112.
33. J.A. Inzana, D. Olvera, S.M. Fuller, J.P. Kelly, O.A. Graeve, E.M. Schwarz et al., 3D printing of composite calcium phosphate and collagen scaffolds for bone regeneration, *Biomaterials* 35 (2014), pp. 4026–4034.
34. T. Billiet, E. Gevaert, T. De Schryver, M. Cornelissen and P. Dubruel, The 3D printing of gelatin methacrylamide cell-laden tissue-engineered constructs with high cell viability, *Biomaterials* 35 (2014), pp. 49–62.
35. K. Tappa and U. Jammalamadaka, Novel biomaterials used in medical 3d printing techniques, *J. Funct. Biomater.* 9 (2018), pp. 17.
36. F. Pati, D.H. Ha, J. Jang, H.H. Han, J.W. Rhie and D.W. Cho, Biomimetic 3D tissue printing for soft tissue regeneration, *Biomaterials* 62 (2015), pp. 164–175.
37. S. Yang, K.F. Leong, Z. Du and C.K. Chua, The design of scaffolds for use in tissue engineering. Part II. Rapid prototyping techniques, *Tissue Eng.* 8 (2002), pp. 1–11.
38. K. von der Mark and J. Park, Engineering biocompatible implant surfaces, *Prog. Mater. Sci.* 58 (2013), pp. 327–381.
39. D. Banoriya, R. Purohit and R.K. Dwivedi, Advanced application of polymer based biomaterials, *Mater. Today Proc.* 4 (2017), pp. 3534–3541.
40. Niinomi M. Mechanical properties of biomedical titanium alloys. Materials Science and Engineering: A. (1998), 243(1-2): 231–6.
41. L. Zhao, P.K. Chu, Y. Zhang and Z. Wu, Antibacterial coatings on titanium implants, *J. Biomed. Mater. Res. Part B Appl. Biomater.* 91B (2009), pp. 470–480.
42. J. An, J.E.M. Teoh, R. Suntornnond and C.K. Chua, Design and 3D printing of scaffolds and tissues, *Engineering* 1 (2015), pp. 261–268.
43. A. Ovsianikov, M. Gruene, M. Pflaum, L. Koch, F. Maiorana, M. Wilhelmi et al., Laser printing of cells into 3D scaffolds, *Biofabrication* 2 (2010), pp. 014104.
44. M. Niinomi, M. Nakai and J. Hieda, Development of new metallic alloys for biomedical applications, *Acta Biomater.* 8 (2012), pp. 3888–3903.

45. J. Serra, P. González, S. Liste, C. Serra, S. Chiussi, B. León et al., FTIR and XPS studies of bioactive silica based glasses, *J. Non. Cryst. Solids* 332 (2003), pp. 20–27.
46. K. Kajihara, Recent advances in sol–gel synthesis of monolithic silica and silica-based glasses, *J. Asian Ceram. Soc.* 1 (2013), pp. 121–133.
47. M. Vallet-Regí and A.J. Salinas, Sol–Gel silica-based biomaterials and bone tissue regeneration, in *Handbook of Sol–Gel Science and Technology*, Springer International Publishing, Cham, 2018, pp. 3597–3618.
48. S. Sultan, G. Siqueira, T. Zimmermann and A.P. Mathew, 3D printing of nano-cellulosic biomaterials for medical applications, *Curr. Opin. Biomed. Eng.* 2 (2017), pp. 29–34.
49. B.K. Gu, D.J. Choi, S.J. Park, M.S. Kim, C.M. Kang and C.-H. Kim, 3-dimensional bioprinting for tissue engineering applications, *Biomater. Res.* 20 (2016), pp. 12.
50. D. Chimene, K.K. Lennox, R.R. Kaunas and A.K. Gaharwar, Advanced Bioinks for 3D Printing: A Materials Science Perspective, *Ann. Biomed. Eng.* 44 (2016), pp. 2090–2102.
51. M. Singh, H.M. Haverinen, P. Dhagat and G.E. Jabbour, Inkjet printing-process and its applications, *Adv. Mater.* 22 (2010), pp. 673–685.
52. P. Calvert, Inkjet printing for materials and devices, *Chem. Mater.* 13 (2001), pp. 3299–3305.
53. C. Prakash, H.K. Kansal, B.S. Pabla and S. Puri, To optimize the surface roughness and microhardness of β-Ti alloy in PMEDM process using Non-dominated Sorting Genetic Algorithm-II, in *2015 2nd International Conference on Recent Advances in Engineering and Computational Sciences, RAECS 2015*, 2016.
54. C. Prakash, S. Singh, S. Sharma, H. Garg, J. Singh, H. Kumar et al., Fabrication of aluminium carbon nano tube silicon carbide particles based hybrid nano-composite by spark plasma sintering, *Mater. Today Proc.* 21 (2020), pp. 1637–1642.
55. C. Prakash, S. Singh, C.I. Pruncu, V. Mishra, G. Królczyk, D.Y. Pimenov et al., Surface modification of Ti-6Al-4V alloy by electrical discharge coating process using partially sintered Ti-Nb electrode, *Materials (Basel)* 12 (2019), pp. 1006.
56. S. Chander, S.K. Mishra, P. Chauhan and Ajai, Ice height and backscattering coefficient variability over greenland ice sheets using SARAL radar altimeter, *Mar. Geod.* 38 (2015), pp. 466–476.
57. C. Prakash, S. Singh, I. Farina, F. Fraternali and L. Feo, Physical-mechanical characterization of biodegradable Mg-3Si-HA composites, *PSU Res. Rev.* 2 (2018), pp. 152–174.
58. C. Prakash, H.K. Kansal, B.S. Pabla and S. Puri, Potential of silicon powder-mixed electro spark alloying for surface modification of β-phase titanium alloy for orthopedic applications, *Mater. Today: Proc.*, 4 (2017), pp. 10080–10083.
59. S. Chander, P. Chauhan and Ajai, Variability of altimetric range correction parameters over indian tropical region using JASON-1 & JASON-2 radar altimeters, *J. Indian Soc. Remote Sens.* 40 (2012), pp. 341–356.
60. C. Prakash, S. Singh, A. Basak, G. Królczyk, A. Pramanik, L. Lamberti et al., Processing of Ti50Nb50-xHAx composites by rapid microwave sintering technique for biomedical applications, *J. Mater. Res. Technol.* 9 (2019), pp. 242–252.
61. C. Prakash, H.K. Kansal, B.S. Pabla and S. Puri, Multi-objective optimization of powder mixed electric discharge machining parameters for fabrication of biocompatible layer on β-Ti alloy using NSGA-II coupled with Taguchi based response surface methodology, *J. Mech. Sci. Technol.* 30 (2016), pp. 4195–4204.
62. A. Sharma, A. Babbar, V. Jain and D. Gupta, Enhancement of surface roughness for brittle material during rotary ultrasonic machining, *MATEC Web Conf.* 249 (2018), pp. 01006.
63. M. Kumar, A. Babbar, A. Sharma and A.S. Shahi, Effect of post weld thermal aging (PWTA) sensitization on micro-hardness and corrosion behavior of AISI 304 weld joints, *J. Phys. Conf. Ser.* 1240 (2019), pp. 012078.

3D Bioprinting 161

64. A. Babbar, V. Jain and D. Gupta, Neurosurgical bone grinding, in *Biomanufacturing*, Springer International Publishing, Cham, 2019, pp. 137–155.

65. A. Babbar, A. Sharma, V. Jain and A.K. Jain, Rotary ultrasonic milling of C/SiC composites fabricated using chemical vapor infiltration and needling technique, *Mater. Res. Express* 6 (2019), pp. 085607.

66. R. Baraiya, A. Babbar, V. Jain and D. Gupta, In-situ simultaneous surface finishing using abrasive flow machining via novel fixture, *J. Manuf. Process.* 50 (2020), pp. 266–278.

67. A. Babbar, A. Kumar, V. Jain and D. Gupta, Enhancement of activated tungsten inert gas (A-TIG) welding using multi-component TiO2-SiO2-Al2O3 hybrid flux, *Measurement* 148 (2019), pp. 106912.

68. V. Keriquel, H. Oliveira, M. Rémy, S. Ziane, S. Delmond, B. Rousseau et al., In situ printing of mesenchymal stromal cells, by laser-assisted bioprinting, for in vivo bone regeneration applications, *Sci. Rep.* 7 (2017), pp. 1778.

69. I.T. Ozbolat, Bioprinting scale-up tissue and organ constructs for transplantation, *Trends Biotechnol.* 33 (2015), pp. 395–400.

70. S.J. Hollister, Porous scaffold design for tissue engineering, *Nat. Mater.* 4 (2005), pp. 518–24.

71. H.-W. Kang, S.J. Lee, I.K. Ko, C. Kengla, J.J. Yoo and A. Atala, A 3D bioprinting system to produce human-scale tissue constructs with structural integrity, *Nat. Biotechnol.* 34 (2016), pp. 312–319.

72. B. Dhandayuthapani, Y. Yoshida, T. Maekawa and D.S. Kumar, Polymeric scaffolds in tissue engineering application: A review, *Int. J. Polym. Sci.* 2011 (2011), pp. 1–19.

73. V. Karageorgiou and D. Kaplan, Porosity of 3D biomaterial scaffolds and osteogenesis, *Biomaterials* 26 (2005), pp. 5474–5491.

74. A. Babbar, P. Singh and H.S. Farwaha, Parametric study of magnetic abrasive finishing of UNS C26000 flat brass plate, *Int. J. Adv. Mechatronics Robot.* 9 (2017), pp. 83–89.

75. K. Hölzl, S. Lin, L. Tytgat, S. Van Vlierberghe, L. Gu and A. Ovsianikov, Bioink properties before, during and after 3D bioprinting, *Biofabrication* 8 (2016), pp. 032002.

76. J. Jia, D.J. Richards, S. Pollard, Y. Tan, J. Rodriguez, R.P. Visconti et al., Engineering alginate as bioink for bioprinting, *Acta Biomater.* 10 (2014), pp. 4323–4331.

77. K. Markstedt, A. Mantas, I. Tournier, H. Martínez Ávila, D. Hägg and P. Gatenholm, 3D bioprinting human chondrocytes with nanocellulose–alginate bioink for cartilage tissue engineering applications, *Biomacromolecules* 16 (2015), pp. 1489–1496.

78. T. Ahlfeld, A.R. Akkineni, Y. Förster, T. Köhler, S. Knaack, M. Gelinsky et al., Design and fabrication of complex scaffolds for bone defect healing: Combined 3D plotting of a calcium phosphate cement and a growth factor-loaded hydrogel, *Ann. Biomed. Eng.* 45 (2017), pp. 224–236.

79. Y. Luo, A. Lode, C. Wu, J. Chang and M. Gelinsky, Alginate/nanohydroxyapatite scaffolds with designed core/shell structures fabricated by 3D plotting and in situ mineralization for bone tissue engineering, *ACS Appl. Mater. Interfaces* 7 (2015), pp. 6541–6549.

8 Investigating on the Lapping and Polishing Process of Cylindrical Rollers

Duc-Nam Nguyen and Ngoc Le Chau
Industrial University of Ho Chi Minh City

Thanh-Phong Dao
Ton Duc Thang University

CONTENTS

8.1	Introduction	164
8.2	Fundamental Principle	164
8.3	Experimental Models to Determine Friction Coefficient in Machining	165
	8.3.1 Setup and Conditions for Lapping Process	165
	8.3.2 Setup and Conditions for Polishing Process	166
8.4	Effects of Experimental Conditions on the Friction Coefficients	166
	8.4.1 Lapping Process with SiC Abrasive Slurry	166
	8.4.2 Polishing Process with Al_2O_3 Abrasive Slurry	169
8.5	Experimental Results for Lapping Process	172
	8.5.1 Experimental Setup	172
	8.5.2 Effect of Abrasive Size to Surface Roughness of Cylindrical Roller	173
	8.5.3 Effect of Downforce to Surface Roughness of Cylindrical Roller	173
	8.5.4 Effect of Downforce to Material Removal Rate of Cylindrical Roller	175
	8.5.5 Effect of Downforce to Roundness of Cylindrical Roller	175
8.6	Experimental Results for Polishing Process	177
	8.6.1 Experimental Setup	177
	8.6.2 Effect of Abrasive Size to Surface Roughness of Cylindrical Roller	177
	8.6.3 Effect of Downforce to Surface Roughness of Cylindrical Roller	178
	8.6.4 Effect of Downforce to Material Removal Rate of Cylindrical Roller	178
	8.6.5 Effect of Downforce to Roundness of Cylindrical Roller	180
8.7	Conclusion	182
References		183

8.1 INTRODUCTION

The bearings have been applied in many industrial fields, such as mechanical engineering, electric motors, automation, cars, and motorcycles. It can work effectively under heavy load conditions [1–3]. Traditionally, the roller's surface was fabricated by CNC turning and grinded finishing [4]. This machining process requires a large amount of time to achieve the required surface quality. Besides, the manufacturing environments such as force, velocity, accuracy, and vibrations of the machine greatly affect the efficiency of the conventional cylindrical roller fabrication process. Therefore, the manufacturing cost will be increased, and production yield will be low.

There have been many products machined by double-side lapping and polishing process. This process can significantly improve the surface roughness and profile accuracy of the workpieces [5–7]. The influence of rollers rotation, material of lapping plate, force, and abrasive slurry was studied and optimized for machining process [8–10]. Experimental results showed that the surface roughness and roundness of the workpieces can be achieved about 0.1 and 0.91 µm, respectively [11–13]. Besides, the influence of machining parameters on surface roughness and material removal of workpiece has been studied in the CMP and e-CMP method [14–21]. As a result, the surface roughness and the roundness decreased less than 0.023 and 0.39 µm. In addition, the motion of workpieces was simulated and analyzed in double-side lapping and polishing process [22–24]. Experimental results show that the surface quality of the rollers is significantly improved [25]. Moreover, the ultrasonic techniques were combined with double-side lapping for processing the crowned roller surfaces [26,27]. The ultrasonic vibration system is put above the upper lapping plate, which generates ultrasonic waves to change the pressure of abrasive slurry during machining process. Therefore, the surface roughness of rollers can be achieved about 44.6 nm after 36 minutes machining.

In this chapter, the basic principle of double-side lapping and polishing process is still to be used in machining of cylindrical rollers. The double-side lapping process is proposed for rough and fine lapping. The influence of abrasive concentrate, abrasive size, and the downforce on the surface roughness and material removal mechanism of workpieces were carried out. Then, the polishing experiments were investigated with different machining parameters.

8.2 FUNDAMENTAL PRINCIPLE

The processing experiments and the schematic of double-side lapping are presented in Figure 8.1, where $\omega 1$, $\omega 2$, $\omega 3$, and $\omega 4$ are the rotation speeds of the lower plate, upper plate, condition rings, and rollers. The cylindrical rollers, the carriers, and conditioning rings were controlled and rotated by the frictional contact to the lower plate during lapping and polishing process. Then, the upper plate was rotated, moved downward, and pressed on the cylindrical rollers by air cylinder and compressor. At this time, the cylindrical rollers will be rolled and machined by two lapping plates under effect of abrasive slurry and downforce.

Lapping and Polishing Process

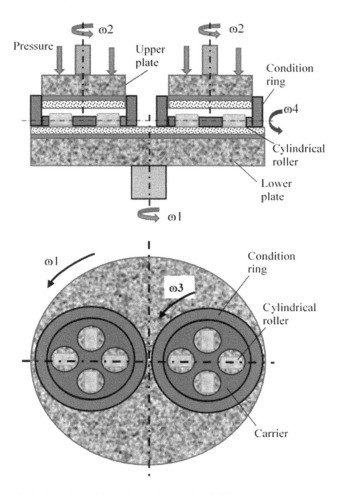

FIGURE 8.1 Principle of double-side lapping and polishing.

8.3 EXPERIMENTAL MODELS TO DETERMINE FRICTION COEFFICIENT IN MACHINING

8.3.1 Setup and Conditions for Lapping Process

The AISI 52100 steel cylindrical rollers (diameter of 15 mm and length of 25 mm) were generated by the turning technique. The experimental results presented that the surface roughness Ra is about 0.5 ± 0.02 μm. Then, the workpieces were machined on the double-side lapping process.

The rolling motion of roller occurs at surface contact of the rollers with the lower plate. Most of the efficiency of machining process depends on the friction force between cylindrical rollers and lower plate. The friction force at contact point A and B was F_{RA} and F_{RB}, respectively. Figure 8.2 shows the friction force of cylindrical

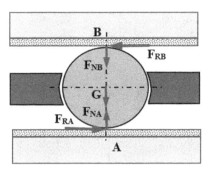

FIGURE 8.2 Schematic diagram of friction forces at contact points.

rollers at contact points. Assuming that the friction coefficients at contact point A and B are equal, the relationship between F_{RA} and F_{RB} is expressed as

$$F_{RA} = F_{NA} \cdot \mu = (F_{NB} + G) \cdot \mu > F_{RB} \tag{8.1}$$

When the pressure force and friction coefficients are increased, the value of this friction force will be increased. Generally, in the lapping process, the contact zone between the workpiece and the lower plate is fed continuously with abrasive grain slurry. The friction coefficient of machining process was tested with a pin on disc apparatus as illustrated in Figure 8.3. With the friction coefficient follows directly as

$$\mu = F_{RA}/F_{NA} \tag{8.2}$$

The normal load was chosen from 15 to 45 N during the experimental process.

The both-side lapping plate with SiC abrasive slurry is used for lapping process of the cylindrical rollers to improve the efficiency of the machining process. The lapping parameters in machining process are listed in Table 8.1.

8.3.2 Setup and Conditions for Polishing Process

The lapping trace and microcrack layer in lapping process will be removed from the workpiece surface by polishing. This process is also carried out with the double-side polishing device. The polishing parameters are listed in Table 8.2, in which the abrasive is changed from SiC to Al_2O_3 in the machining process.

8.4 EFFECTS OF EXPERIMENTAL CONDITIONS ON THE FRICTION COEFFICIENTS

8.4.1 Lapping Process with SiC Abrasive Slurry

The contact zone between the workpiece and the lower plate is fed continuously with abrasive grain slurry. The effects of the abrasive concentration, abrasive size, and the downforce on the friction coefficients were first investigated. Figure 8.3 illustrates a pin on the disc device.

Lapping and Polishing Process

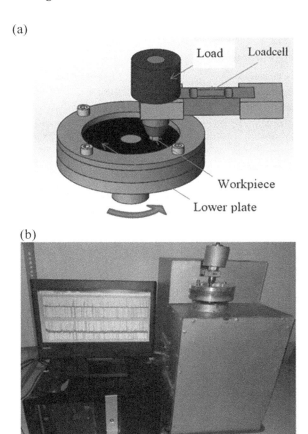

FIGURE 8.3 Picture of the pin on disc device. (a) A custom pin-on-disc friction measuring head; (b) The parts of pin-on-disc friction measuring device.

TABLE 8.1
Lapping Parameters

Workpiece dimensions	Φ15 mm×25 mm
Workpiece material	52100 steel
Lower plate	Cast iron
Abrasive	SiC
Abrasive sizes	800#, 1200#, 4000#
Abrasive concentration (wt.%)	10, 17, 25, 35
Rotating speed (rpm)	100
Lapping time (min.)	2
Downforce (N)	15, 20, 25, 30, 35, 40, 45

TABLE 8.2
Polishing Parameters

Workpiece dimensions	Φ15 mm×25 mm
Workpiece material	52100 steel
Lower plate	Cast iron attached with flannelette
Abrasive	Al_2O_3
Abrasive sizes	1000#, 2000#, 4000#
Abrasive concentration (wt.%)	10, 17, 25, 35
Rotating speed (rpm)	100
Polishing time (min)	2
Downforce (N)	15, 20, 25, 30, 35, 40, 45

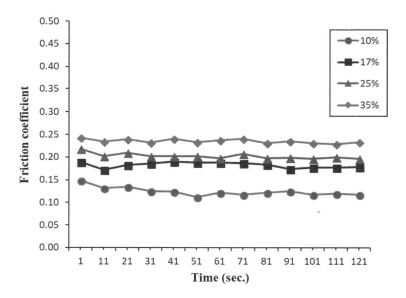

FIGURE 8.4 Relationship between the abrasive concentration and friction coefficients.

Figure 8.4 shows the effect of abrasive concentration on friction coefficients. In the time of 0–2 min, the SiC abrasive size and load were 800# and 20 N. The friction coefficients in this case are about 0.14, 0.18, 0.19, and 0.22 with abrasive concentration of 10%, 17%, 25%, and 35%, respectively.

Based on the experimental results, there has been a slight increase in the friction coefficient value when the abrasive concentration is changed. However, it seems that there is no significant difference in friction coefficients by using abrasive concentration of 17% and 25%. The friction coefficient has reached the maximum value of about 0.22 with abrasive concentration of 35%.

Lapping and Polishing Process

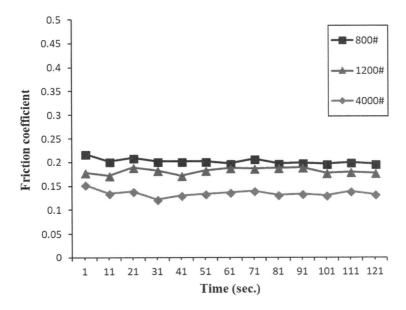

FIGURE 8.5 Effect of abrasive size on friction coefficients.

The abrasive sizes of 800#, 1200#, and 4000# SiC were used in machining process to explore the influence of abrasive grain on the quality of the machining surface. The downforce and abrasive concentration set at 20 N and 25%, respectively.

As shown in Figure 8.5, it can be seen that the change of friction coefficient has little change. The friction coefficient can reach $\mu=0.20$, $\mu=0.19$, and $\mu=0.13$ when the abrasive sizes are 800#, 1200#, and 4000#, respectively. These results indicate that the abrasive size has effect on the friction coefficient.

The relationship between downforce and friction coefficient is presented in Figure 8.6. The abrasive size and abrasive concentration are chosen 800# and 25%, respectively.

From Figure 8.6, it can be seen that seven curves of the friction coefficient are very similar. In the first 10s, the friction coefficient has little change when downforce is changed. There has been a slight increase in the friction coefficient value. In the next time, the friction coefficient was almost unchanged. These results indicate that the downforce has no significant effect on the friction coefficient in the same abrasive slurry. The average friction coefficient in this case is about 0.2.

8.4.2 Polishing Process with Al_2O_3 Abrasive Slurry

The operational precision and working life of rolling bearings depends on the surface quality and profile accuracy of cylindrical rollers. The polishing process with double-side plate is used to achieve the good surface quality. The effects of the abrasive concentration, abrasive size, and the downforce on the friction coefficients were explored in this process.

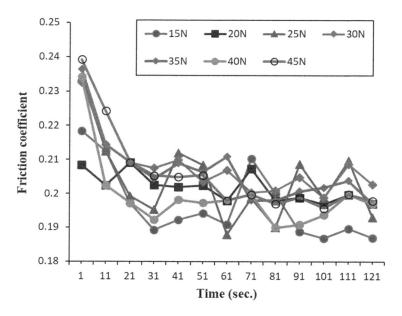

FIGURE 8.6 Relationship between the downforce and friction coefficients.

Figure 8.7 shows the relationship between the abrasive concentration and friction coefficients with Al_2O_3 abrasive slurry. In the time of 0–2 min, the abrasive size and load were chosen of 2000# and 30 N. The friction coefficients in this case are about 0.06, 0.11, 0.13, and 0.16 with abrasive concentrations of 10%, 17%, 25%, and 35%, respectively. The friction coefficient has reached the maximum value of about 0.16 with abrasive concentration of 35%.

The effect of abrasive size on friction coefficient is presented in Figure 8.8. The 1000#, 2000#, and 4000# Al_2O_3 abrasives were chosen for the experiment with the downforce set at 30 N and the abrasive concentration of 25%.

As shown in Figure 8.7, it can be seen that the friction coefficient has little change. The friction coefficient can reach $\mu=0.15$, $\mu=0.13$, and $\mu=0.12$ when the abrasive sizes are 1000#, 2000#, and 4000#, respectively. As a result, the Al_2O_3 abrasive size has little significant effect on the friction coefficient.

The relationship between downforce and friction coefficient is also shown in Figure 8.9. The abrasive size and abrasive concentration are chosen 4000# and 25%, respectively.

Based on Figure 8.9, it can be seen that seven curves of the friction coefficient have changed. The friction coefficient can reach $\mu=0.06$ when the downforce is 15 N. There has been a slight increase in the friction coefficient value when downforce is increased from 15 to 25 N. The friction coefficient in this case is $\mu=0.1$. The maximum value of friction coefficient is 1.6 with the force applied about 45 N. In the force range of 35–45 N, the friction coefficient is almost the same. Therefore, the downforce has greatly affected the friction coefficient in the same abrasive slurry.

Lapping and Polishing Process

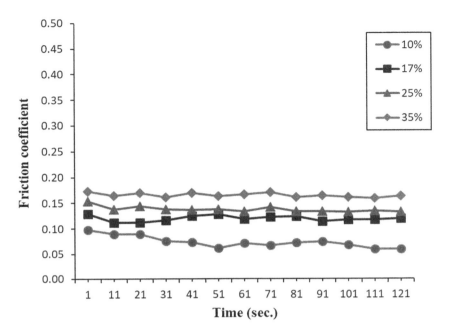

FIGURE 8.7 Relationship between the abrasive concentration and friction coefficients.

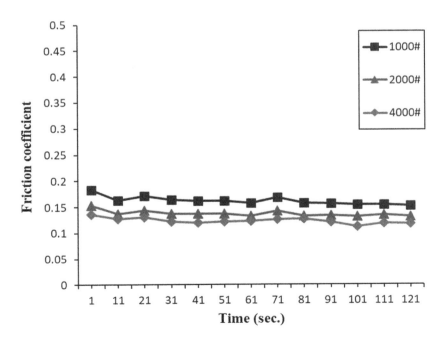

FIGURE 8.8 Relationship between the abrasive size and friction coefficients.

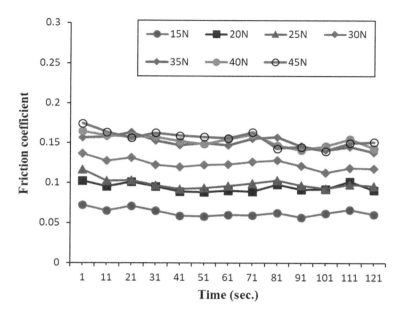

FIGURE 8.9 Relationship between the downforce and friction coefficients.

8.5 EXPERIMENTAL RESULTS FOR LAPPING PROCESS

8.5.1 Experimental Setup

The experiment was carried out on a double-side lapping and polishing device. The cylindrical rollers, the carriers, and conditioning rings were controlled and rotated by the frictional contact to the lower plate during lapping and polishing process (as shown in Figure 8.10). Then the upper plate will freely rotate under load over roller's surface with abrasive particles suspended in water between them. The roller's surfaces were machined via contact between double-side lapping plate and abrasive slurry.

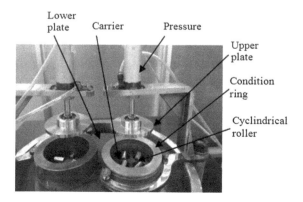

FIGURE 8.10 The double-side lapping device.

Lapping and Polishing Process

TABLE 8.3
Lapping Parameters

Workpiece dimensions	$\Phi15$ mm $\times25$ mm
Workpiece material	52100 steel
Number of workpiece	8
Carrier shape	4 square slots
Lower and upper plate	Cast iron
Abrasive	SiC
Abrasive sizes	800#, 1200#, 4000#
Abrasive concentration (wt.%)	25
Rotating speed (rpm)	150
Lapping time (min)	180
Downforce (N)	15, 20, 25, 30, 35, 40,45

The cylindrical rollers and rigid iron lower and upper plates are driven and rotated in SiC abrasive slurry. The lapping parameters are listed in Table 8.3.

The material removal rate (MRR) of the cylindrical roller in lapping process was measured by a Denver Instrument SI-234 electronic balance. The circular profile and surface roughness of roller was measured with a Rondcom 41C and SJ-401 surface roughness instrument.

8.5.2 EFFECT OF ABRASIVE SIZE TO SURFACE ROUGHNESS OF CYLINDRICAL ROLLER

In order to explore the effect of abrasive size on the surface roughness, the experiment was carried out with the speed, load, and abrasive concentration set at 150 rpm, 30 N, and 25%, respectively. The surface roughness of the workpieces decreases rapidly during the first hours. During the following 3 hours, the surface quality of workpiece has a variation (as presented in Figure 8.11). The roughness can be achieved Ra=0.18 µm, Ra=0.11 µm, and Ra=0.052 µm when the abrasive sizes are 800#, 1200#, and 4000#, respectively. These results indicate that the abrasive size has effect on the surface roughness of cylindrical rollers. The surface quality of the workpieces is reduced from Ra=0.5 µm to Ra=0.052 µm after lapping process.

8.5.3 EFFECT OF DOWNFORCE TO SURFACE ROUGHNESS OF CYLINDRICAL ROLLER

In this section, the experiment is carried out with different downforce values during an hour of lapping process with the SiC slurry. The rotating speed and the abrasive concentration of SiC slurry were at 150 rpm and 25%, respectively. The downforce values of 15, 20, 25, 30, 35, 40, 45, and 50 N were selected for the experiment process. The influence of different force values on the surface roughness is shown in Figure 8.12.

From Figure 8.12, the downforce also had an apparent effect on the surface roughness of cylindrical roller. The surface roughness of rollers is decreased when

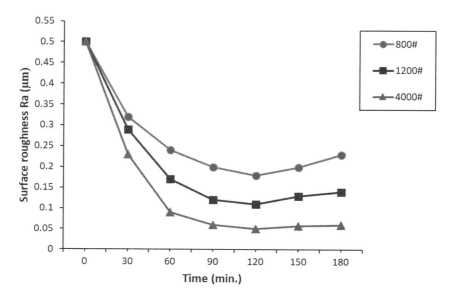

FIGURE 8.11 Effect of abrasive size on surface roughness.

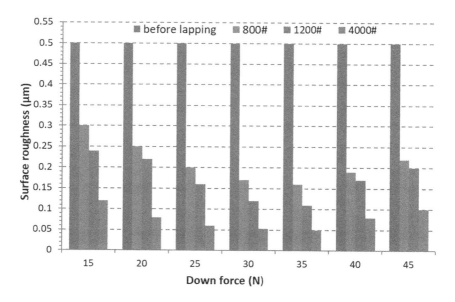

FIGURE 8.12 Effect of downforce on surface roughness.

downforce is increased. A good surface was obtained under condition of the downforce from 30 to 35 N. However, the load exceeds the range value above, and the surface roughness tends to increase. As a result, the new scratches are created on the roller's surface.

Lapping and Polishing Process

8.5.4 Effect of Downforce to Material Removal Rate of Cylindrical Roller

The influence of the downforce on the material removal rate (MRR) of the cylindrical roller is shown in Figure 8.13. The rotating speed and the abrasive concentration of SiC slurry were 150 rpm and 25%, respectively. The different downforce values of 15, 20, 25, 30, 35, 40, 45, and 50 N were set up in the experiment, and the slurry flow rate was chosen of 100 mL/min for experiments.

In the machining time from 0 to 20 min, using the 800# SiC abrasive slurry, the MRR increases from 2.57 to 8.40 μm/min; using 1200# SiC abrasive, the MRR increases from 1.82 to 5.14 μm/min; and using 4000# SiC abrasive, the MRR increases from 1.27 to 4.06 μm/min when the force increases from 15 to 45 N. Based on the experimental results, the downforce and abrasive size have significant effect on the MRR of the cylindrical roller.

8.5.5 Effect of Downforce to Roundness of Cylindrical Roller

The roundness of the cylindrical roller under different conditions of downforce was carried out with SiC abrasive concentration of 25%. Figure 8.14 shows the influence of the downforce on roundness of the cylindrical rollers in lapping process.

From Figure 8.14, it can be seen that the medium roundness of roller has reduced from 3.62 to 2.24 μm with the abrasive size of 800# SiC and downforce of 25 N. When abrasive sizes are 2000# and 4000#, the curves of the roundness are very

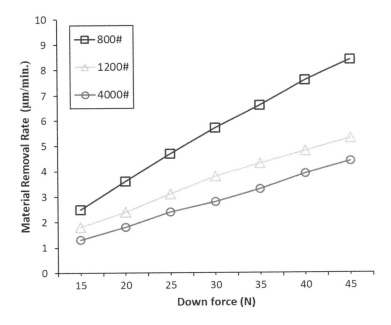

FIGURE 8.13 Effect of downforce on material removal rate.

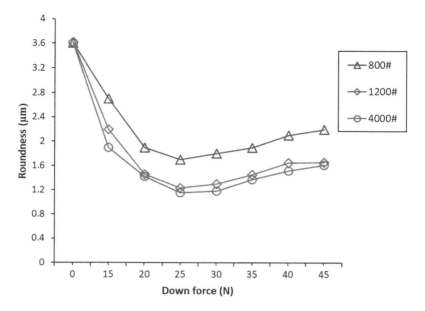

FIGURE 8.14 Effect of downforce on roundness.

similar at different downforce values. The smallest roundness can be achieved about 1.12 µm when the abrasive size and downforce are 4000# and 25 N, respectively.

This roundness will be increased when the downforce exceeds 25 N. Therefore, it can be concluded that the lapping process with different abrasive size is highly effective for the roundness and circular profile of the workpiece. The measured results are presented in Figure 8.15.

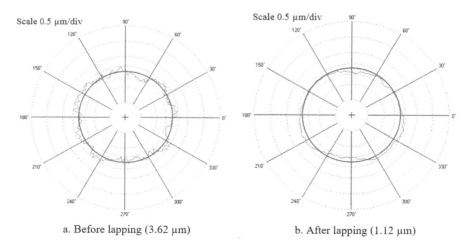

a. Before lapping (3.62 µm)　　　　　　b. After lapping (1.12 µm)

FIGURE 8.15 Measured profiles of cylindrical rollers.

Lapping and Polishing Process

8.6 EXPERIMENTAL RESULTS FOR POLISHING PROCESS

8.6.1 Experimental Setup

The lapping trace and microcrack layer after lapping process will be removed from the workpiece surface by polishing. The outside of two polishing disc will be attached with flannelette pad to restrict the creation of scratches on the roller surface. The experimental setup for the polishing processes is illustrated in Figure 8.16.

In order to explore the effect of abrasive size, the 1000#, 2000#, and 4000# Al_2O_3 abrasives were selected for the experiment with the polishing plate speed set at 100 rpm and the abrasive concentration of 25%. The polishing process is also performed on the double-side polishing device. The polishing parameters are listed in Table 8.4, in which the abrasive is changed from SiC to Al_2O_3 in the fine polishing process.

8.6.2 Effect of Abrasive Size to Surface Roughness of Cylindrical Roller

The effect of abrasive size on surface quality in this process is presented in Figure 8.17. The downforce of 35 N was chosen for the experiment.

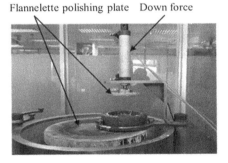

FIGURE 8.16 Schematic of the polishing processes.

TABLE 8.4
Polishing Parameters

Workpiece dimensions	Φ15 mm × 25 mm
Workpiece material	52100 steel
Number of workpiece	8
Carrier shape	4 square slots
Polishing plate	Attached with flannelette
Abrasive	Al_2O_3
Abrasive size	1000#, 2000#, 4000#
Abrasive concentration (wt.%)	25
Rotating speed (rpm)	100
Polishing time (min)	60
Downforce (N)	15, 20, 25, 30, 35, 40, 45

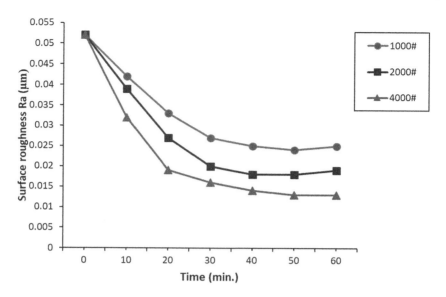

FIGURE 8.17 Effect of abrasive size on surface roughness.

As shown in Figure 8.17, the surface roughness of cylindrical roller will increase when the abrasive size decreases. In the 25 minutes polishing, the surface quality of the rollers decreases rapidly from Ra=0.052 μm to Ra=0.013 μm when abrasive size is 4000# Al_2O_3. In the case of the processing time exceeding 25 minutes, the surface roughness can be achieved Ra=0.025 μm, Ra=0.019 μm, and Ra=0.013 μm when the abrasive sizes are 1000#, 2000#, and 4000#, respectively.

8.6.3 Effect of Downforce to Surface Roughness of Cylindrical Roller

The experiment is carried out with different downforce values during an hour of polishing process with the Al_2O_3 slurry. The rotating speed and abrasive concentration were 100 rpm and 25%. For the experimental process, the downforce values of 15, 20, 25, 30, 35, 40, 45, and 50 N were chosen. The influence of downforce values on surface quality of rollers is shown in Figure 8.18.

From Figure 8.18, downforce is also key factor that influences the surface quality of rollers. The surface roughness of workpieces will decrease rapidly under condition of the downforce from 30 to 35 N. However, the downforce value is not in this range, and the surface roughness tends to increase. The surface quality of rollers can be achieved the minimum value when the applied force is 35 N.

8.6.4 Effect of Downforce to Material Removal Rate of Cylindrical Roller

The influence of the downforce on the MRR of the cylindrical roller is presented in Figure 8.19. The rotating speed of polishing plate and the abrasive concentration of Al_2O_3 slurry was at 100 rpm and 25%, respectively. The different downforce values

Lapping and Polishing Process

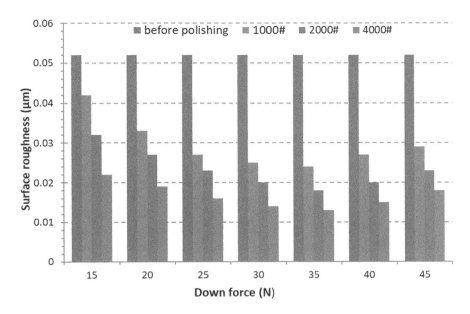

FIGURE 8.18 Relationship between downforce and surface roughness.

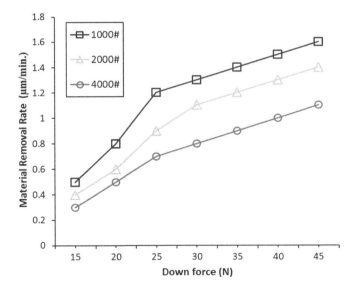

FIGURE 8.19 Relationship between downforce and material removal rate.

of 15, 20, 25, 30, 35, 40, 45, and 50 N were set up in the experiment, and the abrasive slurry flow rate was chosen of 100 mL/min for experiments.

In the time of 0–30 min, using the 1000# Al_2O_3 abrasive slurry, the MRR increases from 0.57 to 1.60 µm/min; using 2000# Al_2O_3 abrasive, the MRR increases from

0.42 to 1.34 μm/min; and using 4000# Al$_2$O$_3$ abrasive, the MRR increases from 0.37 to 1.06 μm/min when the downforce increases from 15 to 45 N. These results indicate that the abrasive size and downforce have little significant effect on the MRR.

8.6.5 Effect of Downforce to Roundness of Cylindrical Roller

The roundness of the cylindrical roller under different conditions of downforce was carried out with Al$_2$O$_3$ abrasive concentration of 25%. Figure 8.20 shows the influence of the downforce on roundness of the workpieces in machining process.

From Figure 8.20, it can be seen that the medium roundness of roller has reduced from 1.62 to 0.93 μm with the abrasive size of 1000# and downforce of 35 N. In addition, the roundness of roller is also decreased from 1.62 to 0.8 and 0.51 μm when the Al$_2$O$_3$ abrasive sizes are 2000# and 4000#, respectively. The best roundness values can be achieved about 0.51 μm when the Al$_2$O$_3$ abrasive size and downforce are 4000# and 35 N.

This roundness will be increased when the downforce exceeds 35 N. The measured profile of workpiece is illustrated in Figure 8.21.

The surface topography was scanned with an MDS metallurgical microscope. Figure 8.22 shows the representative surface topography of the roller's surface in lapping and polishing process. From Figure 8.22b, there are more criss-crossing and scratches on the surface of cylindrical roller after lapping process with #4000 SiC slurry. However, the criss-crossing and tiny scratches of cylindrical surface are reduced in lapping process with 4000# Al$_2$O$_3$ abrasive (as shown in Figure 8.22c).

The images of roller's surface before and after polishing are illustrated in Figure 8.23.

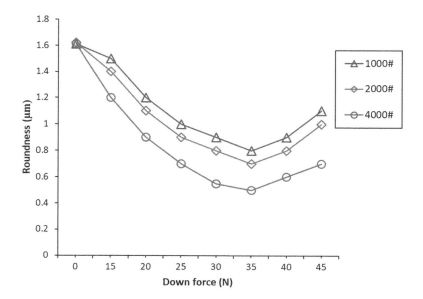

FIGURE 8.20 Relationship between downforce and roundness.

Lapping and Polishing Process

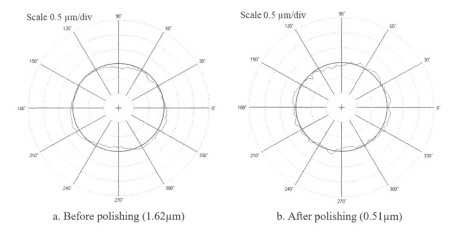

FIGURE 8.21 Measured profiles of cylindrical rollers in polishing process.

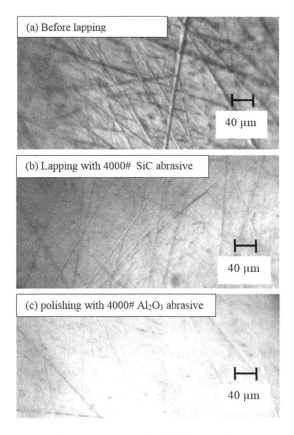

FIGURE 8.22 Typical surface topography of cylindrical surface.

a. Before polishing process

b. After polishing process

FIGURE 8.23 The surface of cylindrical rollers.

8.7 CONCLUSION

In this chapter, the lapping process with SiC abrasive and polishing with flannelette plate are proposed for machining the cylindrical rollers to improve the surface roughness and the roundness of workpiece. The conclusions are shown as follows:

1. From the experimental results, there has been a slight increase in the value of the friction coefficient when the abrasive size is changed. The friction coefficient can reach $\mu=0.20$, $\mu=0.19$, and $\mu=0.13$ when the abrasive sizes of SiC is 800#, 1200#, and 4000#, respectively. However, the friction coefficient has little change when changing machining conditions using Al_2O_3 abrasive slurry.
2. The best surface roughness Ra of 0.052 μm was achieved when using the 4000# SiC abrasive slurry and the downforce of 30 N. However, the best surface roughness Ra of 0.013 μm was achieved when using the 4000# Al_2O_3 abrasive.
3. During the lapping process, using 800# SiC abrasive slurry, the MRR increases from 2.57 to 8.40 μm/min; using 1200# SiC abrasive, the MRR increases from 1.82 to 5.14 μm/min; and using 4000# SiC abrasive, the MRR increases from 1.27 to 4.06 μm/min when the downforce increases from 15 to 45 N.
4. In the polishing process, with 2000# Al_2O_3 abrasive, the MRR increases from 0.57 to 1.60 μm/min; with 2000# Al_2O_3 abrasive, the MRR increases from 0.42 to 1.34 μm/min; and with 4000# Al_2O_3 abrasive, the MRR increases from 0.37 to 1.06 μm/min when the downforce increases from 15

Lapping and Polishing Process

to 45 N. These results indicate that the abrasive size and downforce have little significant effect on the MRR.

5. In polishing process, the roundness of rollers was reduced from 1.62 to 0.93 μm with the Al_2O_3 abrasive size of 1000# and downforce of 35 N. In addition, the roundness of roller is also decreased from 1.62 to 0.8 and 0.51 μm when the Al_2O_3 abrasive sizes are 2000# and 4000#, respectively. The best roundness values can be achieved about 0.51 μm when the Al_2O_3 abrasive size and downforce are 4000# and 35 N.

REFERENCES

1. T.A. Harris, M.N. Kotzalas, *Advanced Concepts of Bearing Technology: Rolling Bearing Analysis.* CRC Press (2006), Taylor & Francis Group.
2. S.M. Ji, F.Q. Xiao, D.P. Tan, Analytical method for softness abrasive flow field based on discrete phase model. *Sci China - Technol Sci.*, 53(10) (2010) 2867–2877.
3. D.P. Tan, S.M. Ji, Y.Z. Fu, An improved soft abrasive flow finishing method based on fluid collision theory. *Int J Adv Manuf Technol.*, 85(5–8) (2016) 1261–1274.
4. F. Hashimoto, I. Gallego, J.F.G. Oliveira, D. Barrenetxea, M. Takahashi, K. Sakakibara, H.O. Stalfelt, G. Staadt, K. Ogawa, Advances in centerless grinding technology. *CIRP Ann Manuf Technol.*, 61(2) (2012) 747–770.
5. M. Song, W. Zou, Improvement of large cylindrical roller processing technology. *J Harbin Bearing*, 35(2) (2014) 65–69.
6. Y. Wang, Q. Deng, L. Cheng, Y. Lv, J. Yuan, Generalization of cylindrical surface processing of cylindrical roller. *Light Ind Mach.*, 30(3) (2012) 110–113.
7. W.H. Zhou, W.F. Yao, M. Feng, B.H. Lv, Q.F. Deng, The polishing process of cylindrical rollers by using a double-side lapping machine. *Key Eng Mater.*, 589–590 (2014) 447–450.
8. K. Nakayama, S. Hashimoto, Experimental investigation of the superfinishing process. *Wear*, 185 (1995) 173–182.
9. S.H. Chang, T.N. Farris, S. Chandrasekar, Experimental characterization of superfinishing. *Proc Inst Mech Eng B-J Eng Manuf.*, 217 (2003) 941–951.
10. Q. Jiang, Z. Ge, Simulation on topography of superfinished roller surfaces. *Sci China Ser B: Chem.*, 45(2) (2002) 122–126.
11. B. Varghese, S. Malkin, Rounding and lobe formation during superfinishing. *J Manuf Process.*, 3(2) (2001) 102–107.
12. S.H. Chang, T.N. Farris, S. Chandrasekar, Experimental analysis on evolution of superfinished surface texture. *J Mater Process Technol.*, 203(1) (2008) 365–371.
13. K. Miura, T. Yamada, M. Takahashi, H.S. Lee, Application of superfinishing to curved surfaces. *Key Eng Mater.*, 581 (2014) 241–246.
14. W. Xu, Z. Wei, J. Sun, L. Wei, Z. Yu, Surface quality prediction and processing parameter determination in electrochemical mechanical polishing of bearing rollers. *Int J Adv Manuf Technol.*, 63(1–4) (2012) 129–136.
15. Z. Wei, W. Xu, B. Tao, J. Song, Crown shaping technique of bearing raceway by electrochemical mechanical machining. *Int J Electrochem Sci.*, 8(2) (2013) 2238–2253.
16. Y.Y. Lin, S.P. Lo, A study on the stress and nonuniformity of the wafer surface for the chemical–mechanical polishing process. *Int J Adv Manuf Technol.*, 22(5–6) (2003) 401–409.
17. D. Zhao, X. Lu, Chemical mechanical polishing: theory and experiment. *Friction*, 1(4) (2013) 306–326.

18. L. Jiang, Y. He, J. Luo, Chemical mechanical polishing of steel substrate using colloidal silica-based slurries. *Appl Surf Sci.*, 330 (2015) 487–495.
19. X. He, Y. Chen, H. Zhao, H. Sun, X. Lu, H. Liang, Y_2O_3 nanosheets as slurry abrasives for chemical-mechanical planarization of copper. *Friction*, 1(4) (2013) 327–332.
20. K. Qin, B. Moudgil, C.W. Park, A chemical mechanical polishing model incorporating both the chemical and mechanical effects. *Thin Solid Films*, 446(2) (2004) 277–286.
21. T.R. Lin, An analytical model of the material removal rate between elastic and elastic-plastic deformation for a polishing process. *Int J Adv Manuf Technol.*, 32(7–8) (2007) 675–681.
22. J. Yuan, W. Yao, P. Zhao, B. Lyu, Z. Chen, M. Zhong, Kinematics and trajectory of both-sides cylindrical lapping process in planetary motion type. *Int J Mach Tools Manuf.*, 92 (2015) 60–71.
23. W. Yao, J. Yuan, F. Zhou, Z. Chen, T. Zhao, M. Zhong, Trajectory analysis and experiments of both-sides cylindrical lapping in eccentric rotation. *Int J Adv Manuf Technol.*, 88 (2017) 2849–2859.
24. S.T. Li, A mathematical model and numeric method for contact analysis of rolling bearings. *Mech Mach Theory*, 119 (2018) 61–73.
25. L. Jiang, W. Yao, Y. He, Z. Cheng, J. Yuan, J. Luo, An experimental investigation of double-side processing of cylindrical rollers using chemical mechanical polishing technique. *Int J Adv Manuf Technol.*, 82 (2016) 523–534.
26. R. Naseri, K. Koohkan, M. Ebrahimi, F. Djavanroodi, H. Ahmadian, Horn design for ultrasonic vibration-aided equal channel angular pressing. *Int J Adv Manuf Technol.*, 90(5–8) (2017) 1727–1734.
27. M. Zhong, J. Yuan, W. Yao, Z. Chen, K. Feng, Double-curved disc ultrasonic-assisted lapping of precision-machined crowned rollers. *Int J Adv Manuf Technol.*, 97(1–4) (2018) 175–188.

9 NiTi Thin-Film Shape Memory Alloys and Their Industrial Application

Ajit Behera and Patitapabana Parida
NIT- Rourkela

Aditya Kumar
ISM-Dhanbad

CONTENTS

9.1 Historical Background of Shape Memory Alloys 185
9.2 What Is Unique about NiTi Alloy? 187
9.3 Stress–Strain–Temperature Curve of a NiTi 187
9.4 Physical Metallurgy of NiTi Thin Film 188
    9.4.1 Phase Diagram 188
    9.4.2 Martensitic Transformation and Crystallography 191
9.5 Physical Properties of the NiTi Thin Film 194
    9.5.1 Field-Emission Scanning Electron Microscopy 194
    9.5.2 Grazing Incidence X-Ray Diffraction 196
    9.5.3 High-Resolution Transmission Electron Microscopy 197
9.6 Applications of Shape Memory Alloys 200
    9.6.1 Microvalves and Micropumps 201
    9.6.2 Microgripper and Microtweezer 202
    9.6.3 Biomedical Equipment 202
9.7 Advantages and Limitations of NiTi Thin film 202
9.8 Conclusions 204
References 204

9.1 HISTORICAL BACKGROUND OF SHAPE MEMORY ALLOYS

The first information on the shape memory effect (SME) of certain alloys was reported in the 1930s. Swedish physicist A. Olander discovered the pseudoelastic behavior of AuCd alloy in 1932 [1]. AuCd alloy can be plastically deformed in the cold state and can recover its undeformed structure at higher temperatures in heating process. The special characteristic of this material is known as SME, and the alloy showing the same characteristic is called shape memory alloy (SMA). In 1938,

Greninger and Mooradian reported on the presence of martensite phase (CuZn) in copper, which is the most responsible phase for SME. Ten years later, Kurdjumov and Khandros proposed the basic of reversible transformation of martensite in SMA in 1949, and thermoelastic martensite transformation was also observed in other alloys such as InTi and CuZn, which were proved by Chang and Read in 1951 [2]. In the 1960s, Buehler and his colleagues discovered the economical SME that is NiTi alloy in the Naval Ordnance Laboratory, United States, which caused great interest in the area of shape memory materials. The name of this Nitinol alloy is Nitinol, which is the abbreviation of "Nitinol Naval Ordnance Laboratory." Leaf and Wallbom first noticed that NiTi and other phases of NiTi have better SME in one phase [3].

Technically, shape memory materials have the following inherent properties that can be applied to intelligent systems:

1. **Sensing**: Due to thermal, electrical, magnetic, and stress stimuli, shape memory materials quickly respond to certain predetermined environmental changes.
2. **Controlled capacity**: Surrounding incentives must reach a predetermined or critical point before operation can begin.
3. **Actuation**: Materials with shape memory can provide greater movement and huge operating forces.
4. **Adaptivity**: The repeatability and reversibility of phase transitions show significant characteristics.
5. **Recovery**: Material configuration is reversible and can be repeated after most of the training cycle.
6. **Energy storage**: A large amount of energy can be used for energy conversion in thermomechanical, magnetic, electromechanical, and chemical machinery.
7. **Damping**: In terms of their inherent phase changes and microstructures, most shape memory materials have a higher specific attenuation capacity.
8. **Mechanism simplicity**: SMA actuators are able to work in a combination of tension, compression, bending, rotation, or some of these deformation modes without any complicated dimension.
9. **Clean operation**: SMA actuators can be created using completely friction-free mechanisms to avoid the generation of any impurities or debris.
10. **Operating energy**: Materials with shape memory require lower working energy. No higher current density is necessary for working.
11. **Noiseless and spark-free operation**: Due to the friction-free and vibration-free components, the movement during operation is extremely quiet. It also provides completely spark-free processes so that they can work in a highly flammable atmosphere.
12. **High power to weight (or power to volume) ratios**: SMAs provide a higher power/weight ratio for small volumes of less than 100 g, which indicates that SMA is extremely attractive in the manufacture of microactuators.
13. **Smaller bandwidth of heating–cooling**: During the heating and cooling process, a smaller bandwidth was observed so that SMEs could respond more quickly. SMA drives can be heated in a variety of ways, such as by

NiTi Thin-Film Shape Memory Alloys 187

radiative or conductive heating of thermal drives and by induction or resistance heating of electric drives. Resistance heating is widely used because of its fast and uniform reaction. Generally, three different types of heat transfer radiation, conduction, and convection are observed. In general, the effects of radiation can be ignored because the temperature used is below 100°C, and because the encapsulated liquid does not move, it is conductive.

14. **Reliability**: The reliability of SMA equipment depends on its service life worldwide. The most important control parameter for SMA reliability is the conversion period affected by time, temperature, stress, and conversion stress.

15. **Multifunctionality**: SMAs can serve as multifunctional materials by using different training media, such as thermomechanical or magnetomechanical processes.

After sufficient cyclic motion, many SMAs end by generating significant plastic elongation, and this mechanism stabilizes the response of the entire material. The repetition of the cyclic process is called training. For thermally responsive SMA, training is usually performed by maintaining a constant temperature. Different cycles are required before the stress/elongation response can be repeated [4]. Unidirectional and bidirectional SMAs were developed based on the training process.

9.2 WHAT IS UNIQUE ABOUT NiTi ALLOY?

Many studies have been conducted to find out how to choose different materials to make microelectromechanical systems (MEMS) and bio-MEMS. Among the previously discovered SMAs, NiTi SMA has proven to be a greater potential element for MEMS due to its unique SME, high energy output, fairly cheap price, and higher biocompatibility. These important aspects have led to an in-depth study of the SME and its applications. The use of Nitinol is fascinating because Nitinol has unique functional properties compared with other conventional alloys. However, it took almost 20 years from the beginning to the industry to accept it. The first important industrial application is the "rivets" that connect thick steel beams to bridges. That rivet is like a nail, with heads on both sides, to hold the item at the required temperature. But the first Nitinol product was Raychem Corporation's Crychemit™ "shrink-to-fit" fitting, introduced in 1969. Grumman Aerospace used this heat shrink tube and solved the coupling problem of hydraulic oil lines in F-14 fighters [5]. There is another type of SMA, called a ferromagnetic shape memory alloy (FSMA), that changes shape in response to a strong external magnetic field. FSMA is more effective than temperature-induced SMA. The application of NiTi in MEMS began in the early 1990s, when Walker et al. Manufactured SMA coil springs on silicon wafers. SMA films made by various technologies have attracted great interest in the fields of actuators, microvalves, microfluidic pumps, and microgrippers.

9.3 STRESS–STRAIN–TEMPERATURE CURVE OF A NiTi

Figure 9.1 graphically shows the stress–strain temperature (σ-ε-T) curve of NiTi SMA used to illustrate SMEs. At the beginning of the charge path A in Figure 9.1, the SMA is in the austenite parent phase. After cooling to point B, the SMA will

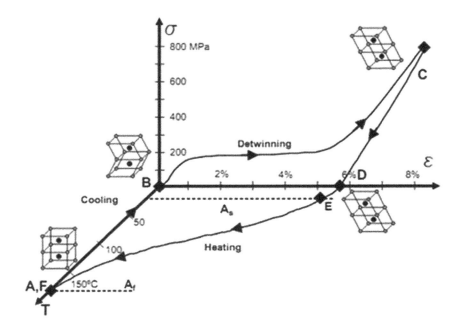

FIGURE 9.1 This graph shows the experimental stress–strain temperature curve (σ-ε-T) of NiTi SMA. SMA, shape memory alloy.

transform to martensite without applying stress and reach an adaptive or merged configuration. Because tension is applied from point B to point C, the recovery from the beginning of die casting causes deformation, and significant macroscopic strain (at point C) is observed. The unloading starts at point C and restores the elastic elongation, whereas the inelastic elongation due to the detinning process is retained due to the stability of the unentangled martensite at point D. During the heating process, the SMA releases all voltage at point E. Here the temperature reaches As, and the reverse transformation to the austenite parent phase begins at point F of the austenite end temperature (Af). The inelastic strain was restored as a result of the redirection, and the original structural form was restored (before deformation B–C). Here, the formation of irrecoverable plastic stretching is ignored, and point A is equal to point F in terms of the state of the material. The enhancement of the original shape of the alloy is called "shape memory alloy." Similarly, cooling without stress will result in no change in the shape of the double martensite in the same way as the loading path.

9.4 PHYSICAL METALLURGY OF NiTi THIN FILM

9.4.1 Phase Diagram

Phase diagrams are critical to understanding how alloys are handled and developed. Phase diagrams can be used to understand the relationship between care transformation temperature and microstructure, composition, and sediment development. It has also been reported that the phase transition temperature plays an important role in SME behavior, which is strongly influenced by film composition, manufacturing conditions, and postdeposition annealing. However, there are many problems in constructing the phase diagram of the NiTi system due to the following reasons: (i) Ti is highly active and will rapidly combine with oxygen, nitrogen, and carbon at high temperatures, and (ii) there are different metastable phases in the medium temperature range, and one of them has a strong impact on the transformation path and characteristics of SMEs. This is why there is currently no complete phase diagram of the NiTi system [6].

Three equilibrium intermetallic phases can be generated in the NiTi system: Ni_3Ti, NiTi, and $NiTi_2$ (Figure 9.2). The Ni_4Ti_3 and Ni_3Ti_2 phases are mesophases, they are transformed into a balanced Ni_3Ti phase, and the aging time is longer [7]. However, Raghavendra reported [8] that according to the annealing temperature and time, Ti_3Ni_4, Ti_2Ni_3, and $TiNi_3$ were precipitated in sequence from the three phases. At lower annealing temperature and shorter soaking time, Ti_3Ni_4 phase will be precipitated, whereas at higher annealing temperature and longer soaking time, $TiNi_3$ phase will be formed, and at intermediate annealing temperature and time, the Ti_2Ni_3 phase will precipitate. In addition, with continuous aging, the previously formed Ti_3Ni_4 is dissolved in the matrix, and the Ti_2Ni_3 phase increases. With further aging, Ti_2Ni_3 is reabsorbed by the matrix and replaced by $Ti Ni_{13}$ precipitates. These observations invalidate many previous comments or suggestions on the properties of the different phases identified in the NiTi alloy system and indicate that Ti_3Ni_4 and Ti_2Ni_3 are only intermediate phases and $TiNi_3$ is the equilibrium phase:

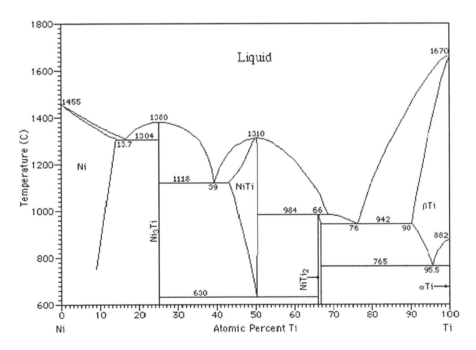

FIGURE 9.2 Ni-Ti phase diagram [16].

$$\text{NiTiB}_2 \text{ parentphase} \rightarrow \alpha_1 + \text{Ni}_4\text{Ti}_3 \rightarrow \alpha_2\text{Ni}_3\text{Ti}_2 \rightarrow \alpha_3\text{Ni}_3\text{Ti}$$

where α_1, α_2, α_3 are matrices with different Ni concentrations.

Taylor and Freud examined the structure of Ni_3Ti and reported a hexagonal, densely packed DO24-type ordered structure with a space group of P63/mmc. The lattice constants are $a = 5.1010$ Å and $c = 8.3067$ Å [9]. The lattice parameter of Ni_3Ti is about twice that of NiTi. Ti_2Ni_3 is the second metastable precipitate and forms a longer aging time at average or low temperatures. Nishida and Wayman [10] reported in situ TEM with heated sample holders for phase transitions associated with Ti_2Ni_3 precipitates in aged $\text{Ni}_{52}\text{Ti}_{15}$ alloys. They reported two main findings: (i) the precipitate showed a two-step transformation; the tetragonal phase became an orthogonal phase at high temperature and finally became a monoclinic phase at low temperature, and (ii) the morphology of these phases is different; the meso-phase of the cuboid is inverted, and the monoclinic phase is needle shaped. Twins apparently caused needle-like area contrast. Hara et al. [11] reported that, relative to temperature, precipitation has two phases, and the phase change occurs through the formation of martensite in the mesophase, similar to the transformation of the R phase, and the transformation of the B2 phase was observed. Go to B19. It was found that the high-temperature phase (stable above 100°C) has a tetragonal structure with a lattice parameter $a = 3.095$ Å and $c = 13.585$ Å (at 100°C); the lower phase has an orthogonal structure with a lattice parameter $a = 4.398$ Å, $b = 4.370$ Å, $c = 13.544$ Å (at 25°C) [11]. However, they did not report the structure of the monoclinic phase at low temperatures. Ti_3Ni_4 is an essential phase because it affects the SMA properties of nickel-rich NiTi alloys [12]. Ni_4Ti_3 is usually in the form of a lenticular lens. Nishida et al. [7] and Saburi et al. [13] calculated the lattice constants of diamond-shaped Ti_3Ni_4 ($a = 6.70$ Å, $a = 113.8$ Å), whereas Tadaki et al. And Saburi et al. calculated the lattice constants. The crystal structure is rhombohedral [13,14]. According to reports, NiTi_2 has an FCC structure, and its lattice parameters are five times larger than NiTi [8]. The NiTi_2 equilibrium phase has a high- and low-temperature crystal structure due to the martensitic transformation of the martensite. But Mr. Wang reported that NiTi_2 precipitates grow uncontrollably, which hinders the formation of martensite plates [15].

The meaning of the phase diagram is (i) the limit for forming NiTi on the titanium-rich side is vertical; (ii) the limit for solubility of NiTi on the nickel-rich side decreases with decreasing temperature, and the solubility is about 50°C negligible; (iii) the possibility of decomposition of the eutectoid at 63°C is unrealistic, as no direct confirmation was found; (iv) the Ni_3Ti_4 phase appears at a lower aging temperature and has a longer aging time; (v) the Ni_3Ti phase appears at a higher aging temperature and has a longer aging time; (vi) the Ni_3Ti_2 phase appears at the average aging temperature; (vii) by increasing the aging time of Ni_4Ti_3 observed in the matrix, the size of the Ni_2Ti_3 phase increases; (viii) by extending the aging time, the existing Ni_3Ti_2 phase found in the matrix was obtained, and the number and size of the Ni_3Ti phases were increased; (ix) it is confirmed that both the Ni_4Ti_3 and Ni_3Ti_2 phases are mesophases, and the diffusive phase transition occurs in the following order with increasing aging temperature and time; and (x) list of all the invariant points on the NiTi phase diagram is given in Table 9.1.

TABLE 9.1
Different Invariant Reaction Points in NiTi System

Temperature (°C)	Different invariant Points
1670	Pure metal melting point (invariant)
882	Pure metal crystal structure change (invariant)
786	Eutectoid
942	Eutectic
984	Peritectic
630	Eutectoid
1310	Congruent melting point
1118	Eutectic
1380	Congruent melting point
1300	Eutectic
1455	Pure metal melting point (invariant)

$$Ni_4Ti_3 \rightarrow Ni_3Ti_2 \rightarrow Ni_3Ti$$

Metallurgical factors such as annealing and aging treatments and annealing time are very sensitive to the phase transformation temperatures. The TTT diagram of $Ni_{52}Ti_{48}$ has been given in Figure 9.3.

9.4.2 Martensitic Transformation and Crystallography

A martensitic transformation is a particular category of solid-state phase transformations. Martensitic transformation is a type of solid-state transformation, which is a shear-dominated, nondiffusive transformation that does not comply with Fick's diffusion law. By maintaining the relative orientation relationship between the atom and its neighbors during the phase transition, the atomic motion is limited to less than the distance between atoms. This type of martensitic transformation can also be observed in nonferrous alloys, pure metals, ceramics, and polymers. All materials undergoing martensitic transformation need not exhibit SMEs. The martensitic phase specifically alters the thermoelastic martensitic transformation to treat the functional behavior shown by SMA. Three types of conversion pathways have been observed in different NiTi-based films:

1. Parent phase (B2 phase, cubic structure) → martensite phase (B19′ phase, M phase, monoclinic structure),
2. B2 phase → B19 phase (O phase, orthorhombic structure) → B19′ phase, and
3. B2 phase→R phase (Trigonal) → B19′ phase.

The second path is due to the effect of additional third element copper. The graphical representation of above three transformation paths of NiTi alloy has been given in Figure 9.4.

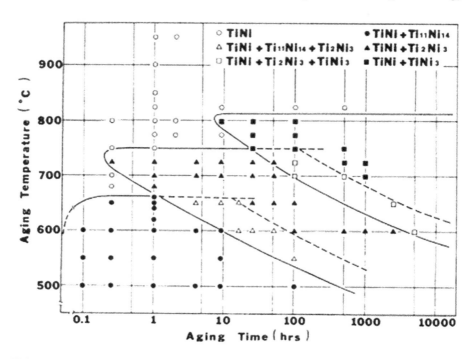

FIGURE 9.3 The TTT diagram is explaining the aging behavior for Ni52Ti48 alloy [3].

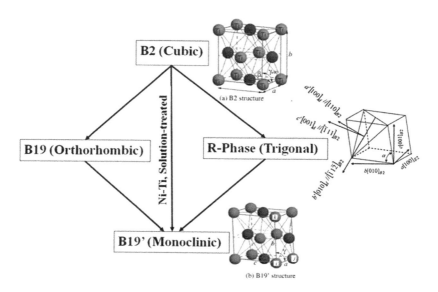

FIGURE 9.4 Three-phase transformation paths in NiTi-based alloys [3].

NiTi Thin-Film Shape Memory Alloys

B19' is the most common martensite phase observed in NiTi and its alloys. It is known here that the structure of B19 has a lower symmetry than B2. That is, SMA has undergone a transition from a cubic austenite phase with high symmetry (high temperature) to a monoclinic martensite phase with low symmetry (low temperature). B19' has a space group of P21/m and has a monoclinic crystal structure. For TiNi49.2, its lattice parameters are $a=0.2889$ nm, $b=0.4108$ nm, and $c=0.4646$ nm and $\beta=97.780$. The mesh parameters depend on the composition. The other phase, called R phase, belongs to the space group of P3, and its lattice parameters are $a=0.732$ nm and $c=0.532$ nm. The R phase extends 0.94% in the direction of the upper [111] B2, which is one order smaller than that of B19'' [1,2]. B19 is an orthogonal crystal phase with a lattice constant of $a=0.29$ nm, $b=0.425$ nm, and $c=0.45$ nm. The B2→R conversion immediately follows and starts from B2+B19'. Usually, we can write the conversion path as B2→ B2+B19'→ R+B19'→ B19', the second phase of which is narrow, observed in the temperature range (60°C–55°C) [17]. Miyazaki believed that the appearance of the R phase was due to precipitation (such as Ti_2Ni in titanium-rich materials or Ni_4Ti_3 in nickel-rich materials) and fine grain structure, which may hinder martensitic transformation [18,19]. Compared with the martensitic transformation, the hysteresis of the R phase transformation is very small in NiTi SMA. Therefore, the R phase transition plays an effective role in improving the response of high-speed microactuators [20]. Due to the deformation of the B2 battery in the <111> direction, the R phase is formed. The transformation from cubic B2 austenite to the diamond phase is initiated by deformation during cooling at the diamond starting temperature. Thermal cycling under no stress conditions is conducive to the formation of the R phase in NiTi. Another condition is the introduction of dislocations and internal stress through thermal cycling that allows R phase nucleation [21]. Various factors that inhibit the onset temperature of martensite are (i) increasing the percentage of Ni, (ii) aging at intermediate temperature, (iii) annealing at recrystallization temperature after cold working, and (iv) replacing the third element and (v) precipitation and fine grain size. Phase transition temperature will (i) inhibit R phase; (ii) increase grain size; and (iii) reduce oxygen pollution.

It is recognized that the three-step transition is related to the heterogeneity of the microstructure at the micro/macroscale of the film. Microscopically, changes in the Ni concentration around metastable Ni_4Ti_3 precipitates suggest coherent fields and subgrain boundaries around these precipitates, leading to multistep martensitic transformations [22]. Similarly, nonuniform precipitation near the grain boundaries results in multistep martensitic transformation. All the possibilities of the three-phase phase transition affected by the nonuniformity in NiTi films are described as follows:

1. Stress distribution: Due to stress inhomogeneity, the R phase transition follows one step, whereas the R→B19 transition follows two steps, corresponding to the high stress area (close to precipitation) and the low stress area (precipitation pathway).

2. Composition inhomogeneity: The heterogeneity of the composition in the NiTi system (between Ni_4Ti_3 particles) will produce a B2→R transition and two R→ B19' transitions, the latter corresponding to the transition in

the low Ni region (near Ni_4Ti_3) and the high Ni region (away from Ni_4Ti_3). Ni_4Ti_3 particles are best to be grown around the grain boundaries and much less inside the grains.

3. Grain boundary effect: There has been observed low Ni content (50.6Ni) and very low nucleation rate, so the precipitation of Ti_3Ni_4 is very sensitive to the presence of grain boundaries.

Ni_4Ti_3 is not an equilibrium phase. It has been found that there is a large difference in the response of the stress–strain behavior due to the internal stress caused by the Ni_4Ti_3 precipitates [23]. The NiTi intermetallic compound melt phase is $NiTi_2$. Low-phase intermetallic compounds have many attractive properties such as high hardness, high melting point, excellent chemical stability, and strong atomic bonding. $NiTi_2$ has a complex, face-centered cubic structure with 96 atoms per unit cell (space group Fd3m). According to reports, due to the high concentration of titanium in the titanium-rich film, $NiTi_2$ nucleation precipitated during the crystallization process and grew along the austenite grain boundaries.

9.5 PHYSICAL PROPERTIES OF THE NiTi THIN FILM

Physical properties of the experimental data on the microstructure, surface and interface morphology, crystal structure, and phase formation of the NiTi thin film were analyzed. Different technologies for thin-film physical characterization are field-emission scanning electron microscope (FESEM), grazing incidence X-ray diffraction (GIXRD), high-resolution transmission electron microscope (HRTEM), and atomic force microscope (AFM).

9.5.1 FIELD-EMISSION SCANNING ELECTRON MICROSCOPY

FESEM is a versatile nondestructive technique that displays detailed information about the morphology of a sample. The field-emission cannon used in FESEM produces sharper, less electrostatically distorted images with a spatial resolution of up to 1.5 nm, which is three to six times that of traditional scanning electron microscopes. Figure 9.5a and b shows the surface morphology and interface micrographs of two NiTi films deposited by sputtering. The surface image shows a tear-free solid surface containing the microstructure of an orange peel island with almost no depressions between the two islands. Here, the atoms are deposited in a columnar manner, as shown in Figure 9.5b. Here, the Ni layer and the Ti layer are both crystalline in nature, and a diffusion amorphous layer having a minimum thickness exists at an interface between the two layers. For the deposited samples, the interface is not sharp, and some Ni and Ti interdiffusion occurred during the deposition process. Ni can diffuse into Ti quickly enough to consume the entire layer before the ternary compound is formed.

Figure 9.6a shows the surface morphology, and Figure 9.6b shows the interface morphology of single-layer and double-layer Ni/Ti films annealed at 600°C. During the annealing process, atomic migration occurs, resulting in an increase in the percentage of porosity along the island boundary. During the pillar deposition process,

NiTi Thin-Film Shape Memory Alloys

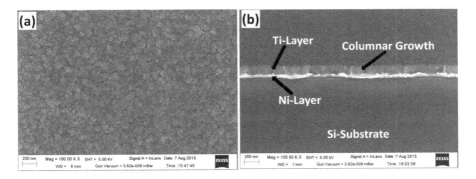

FIGURE 9.5 Sputter-deposited NiTi single bilayered thin film of (a) surface morphology and (b) interface morphology.

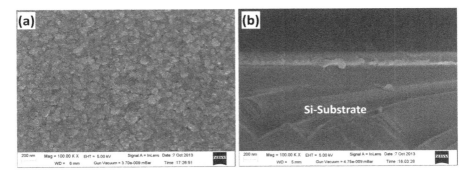

FIGURE 9.6 NiTi single bilayered thin film, after annealing at 600°C of (a) surface morphology and (b) interface morphology.

the partially closed pores below the surface change at high temperatures and become open pores. These openings are not interconnected. Due to the temperature rise, two phenomena have appeared, atomic migration and fusion of atomic clusters, and are closely related to certain processing parameters such as glow time and substrate cooling rate. Diffusion between layers increases with increasing annealing temperature. Once an amorphous phase is formed, its growth requires mutual diffusion through the amorphous phase. If this process is slow, the amorphous/crystalline interface will proceed slowly, allowing more time for nucleation of competing intermetallic connections at these interfaces. At higher annealing temperatures, atomic migration on the surface and diffusion between layers increase, and the film surface looks like a flat surface with lower surface roughness (Figure 9.6a). The annealing temperature is limited to 600°C to prevent oxidation of NiTi films by atmospheric oxygen [24]. At lower annealing temperatures (300°C), induced stresses can cause cracks between the film and the substrate. The initial stress may be due to (i) the molar volume difference between the crystalline and amorphous phases, because the molar volume of the amorphous Ni_xTi_{1-x} is smaller than the mixture of crystalline nickel and titanium [25], (ii) thermal mismatch, and (iii) the coherence of the new phase or precipitation phase. Continuous annealing preferentially leads to greater diffusion, which leads to stress relaxation. At a glow temperature of 600°C, the diffusion of Ni and Ti atoms onto the substrate results in the formation of binary and ternary silicides. The maximum stress values in the nickel and titanium phases are expected to be approximately +1000 and −1100 MPa, respectively [25]. When annealed at 600°C, diffusion is greatest in each layer and looks like a single layer (Figure 9.6b). In the interface microstructure, we observed a gloss difference between the layer and the interface. This is due to the presence of intermetallic phases and composite compounds through interfacial diffusion and chemical interactions. NiTi intermetallic compounds (NiTi, $NiTi_2$, Ni_3Ti, Ni_3Ti_4, Ni_4Ti_3) exist together with binary and ternary silicides on the Ni and Ti layers. Whang et al. pointed out that by ignoring the compaction effect during structure relaxation, it is reasonable to estimate that it is about 1% higher than the corresponding crystalline phase observed in Ni40Ti60 [26]. This means that our sample undergoes volume shrinkage after the transition from crystalline nickel and titanium to amorphous state after annealing [25]. Due to (i) the large mixed negative enthalpy as the driving force of the reaction and (ii) the rapid diffusion of one element into the crystal of the other element, the formation of a metastable amorphous phase is better than a stable crystalline state.

9.5.2 Grazing Incidence X-Ray Diffraction

GIXRD technology is used to detect the presence of crystal phases in deposited and annealed Ni/Ti films. If the angle of incidence is less than the critical angle θc associated with the material, X-rays at the viewing angle will have total external reflection on the surface. Figure 9.7 shows the GIXRD patterns of Ni/Ti films before and after annealing. The formation of NiTi intermetallic phase was observed on the surface of the binary silicide sample. The silicide compound is formed due to the high energy (higher annealing temperature) applied to the substrate, which is sufficient to soften the substrate and diffuse it onto the film. As seen in the sample (Figure 9.7a); NiTi, Ni_3Ti, and $NiTi_2$ are present. In the early stage of interdiffusion, the nucleation and growth kinetics of various possible reaction products can promote the formation of nonequilibrium phases. Recently, amorphous phases formed due to low-temperature solid-state interdiffusion in two metal–metal layers have been reported [27]. The formation of binary silicides results in the formation of ternary silicides during the deposition process. At higher annealing temperatures (600°C) (Figure 9.7e), Ni_4Ti_3, Ni_3Ti, and binary and ternary silicides ($TiSi_2$, Ni_2Si, $Ni_4Ti_4Si_7$) form intermetallic compounds. The surfaces of all the samples are mainly NiTi intermetallic compounds. However, it was found that the observed concentrations of nickel and titanium species changed with the annealing treatment. Thermodynamics promotes the formation of intermetallic compounds rather than amorphous phases. Therefore, the kinetics of nucleation and growth must be responsible for the appearance of the amorphous phase during the solid-state reaction. Formation of the formation can be used as a measure of the thermodynamic driving force for forming different phases at low temperatures. The thermochemical data of the Ni–Ti system estimated by Meng et al. showed that different Ni–Ti compounds of about 35 kJ/mol

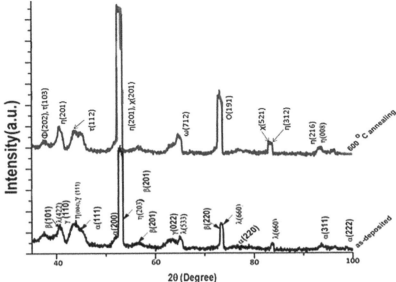

FIGURE 9.7 Combined GIXRD pattern of the Ni/Ti samples: (a) as-deposited and annealed at (b) 600°C. GIXRD, Grazing incidence X-ray diffraction.

were formed [27]. In the early stages of the interdiffusion reaction of thin-film diffusion pairs, sensitive interactions between kinetic and thermodynamic effects are effective. Setton et al. explained about important phase reactions that occur during sample deposition:

$$3Ni + Ti = Ni_3Ti \tag{9.1}$$

$$5Ni + Ti_2Ni = 2TiNi_3 \tag{9.2}$$

$$2Ti + Ni = Ti_2Ni \tag{9.3}$$

$$5Ti + TiNi_3 = 3Ti_2Ni \tag{9.4}$$

9.5.3 High-Resolution Transmission Electron Microscopy

The surface and interface morphology of the deposited and annealed NiTi films were observed with an HRTEM. Figure 9.8a shows the TEM edges of the surface of a single bilayer film of NiTi. The surface morphology showed frequent replacement twins, and the widths of these parallel slats were comparable. Different twinning orientations focus on the heterogeneity of the material. A selected area diffraction (SAD) pattern of the surface is shown in Figure 9.8b. Here, it is clear that the Ni_3Ti_4, Ni_3Ti, Ti, and $NiTi_2$ phases exist at the levels of (131), (200), (101), and (844),

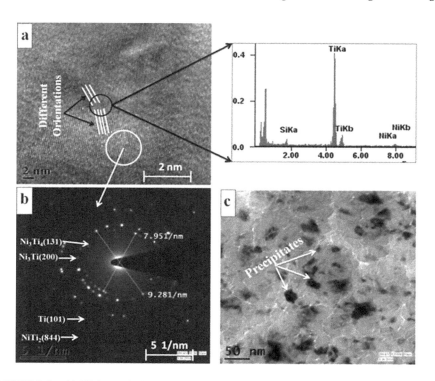

FIGURE 9.8 (a) High-resolution lattice images of as-deposited Ni/Ti surface with EDS. (b) The SAD pattern near the top surface of the Ni/Ti thin film showing several planes on the ring. (c) Bright field image showing the presence of intermetallic precipitates. EDS, energy-dispersive X-ray spectroscopy; SAD, selected area diffraction.

respectively. Figure 9.8c shows a clear field of view (BF) image of the Ni/Ti film. Note that the transition enters the crystalline/amorphous interface. This interface provides different nucleation sites, and every transformation starts at this interface.

Figure 9.9a and b shows interface micrographs of NiTi bilayers with diffusion points along the interface. Figure 9.9c shows the Ni/Ti interface in SAD mode. It indicates that there are some different phases in the reaction products at the interface. Using SAD pattern analysis, NiTi and Ni_3Ti were found at the interface (between the film and the lighter gray ternary connection). The preferential diffusion takes place in the direction of the layers, crossing the interface at an angle, which is beneficial for Ni diffusion (Figure 9.9b). The nickel atom appears to be the main diffusing substance in the substrate. The SAD pattern on the interface (Figure 9.9c) indicates the presence of intermetallic compounds NiTi and Ni_3Ti along the (220) and (213) planes, respectively. The formation and interdiffusion of these phases lead to the amorphization of the Ni/Ti bilayer film at the interface. Amorphous layer thickness increases with increasing annealing temperature.

The surface morphology of Ni/Ti films at high annealing temperature (600°C) is shown in Figure 9.10. This figure shows the grain diffusion and the diffusion of atom clusters on the surface, and various atom orientations are clearly observed in Figure 9.10a and b. The presence of stress surfaces and twisted structures is shown in

NiTi Thin-Film Shape Memory Alloys 199

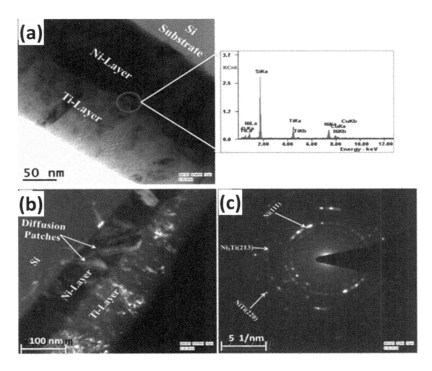

FIGURE 9.9 (a) HRTEM microstructure of the interface of Ni/Ti bilayer thin film on Si substrate. (b) Dark field image indicating diffusion patches. (c) The SAD pattern at the interface of Ni/Ti thin film. HRTEM, high-resolution transmission electron microscope; SAD, selected area diffraction.

Figure 9.10c. Figure 9.10d shows the SAD pattern of the surface. SAD mode analysis has informed us of the existence of NiTi, Ni_4Ti_3, $NiTi_2$, and Ni_3Ti intermetallic phases in the (100), (122), (440), and (105) planes, respectively. The presence of the Ni_2Si binary silicide compound was detected to grow along the (200) plane of the substrate orientation. Here, Ni_2Si forms the earliest of all six stable binary silicides (Ni_3Si, $Ni_{31}Si_{12}$, Ni_2Si, Ni_3Si_2, $NiSi$, and $NiSi_2$) [28]. In addition to the ring corresponding to the unreacted metal, new rings may appear. These new rings are close to the rings of NiTi's B2 (CsCl type) intermetallic compound. This identification is not definitive; slightly different heat treatments cause slight differences in these diffraction rings [27,30].

In the dark field image, different precipitations and phases were observed in this layer. Generally, the Ni atom is the main mobile species in the amorphization reaction of the element Ni/Ti diffusion pair. Ni atoms move in the Ti lattice and leave defects in the Ni lattice, leading to a further diffusion process. We consider the following growth sequence of the amorphous reaction:

Elemental Ti crystal→supersaturated solution of Ni in Ti→interfacial layer of amorphous $Ni_{1-x}Ti_x$→ bulk amorphous $Ni_{1-x}Ti_x$→equilibrium $Ni_{1-x}Ti_x$.

The interdiffusion of Ni in the Ti parent lattice results in an unbalanced concentration of Ni in Ti. The stabilization of the Ti lattice occurs through the formation of

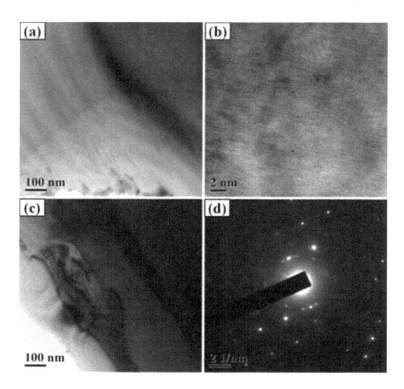

FIGURE 9.10 HRTEM surface micrograph of single bilayered thin-film surface annealed at 600°C (a) showing grain diffusion and atomic cluster diffusion on the surface, (b) different atomic orientations, (c) stresses surface and presence of twined structure, (d) SAD pattern of the surface. Deposition time for each layer is 25 minutes, and annealing time is 1 hour. HRTEM, high-resolution transmission electron microscope; SAD, selected area diffraction.

an amorphous layer at the Ni/Ti interface, leading to phase growth. By increasing the thickness of the amorphous layer, there will be an amorphous phase relaxation in the overall metastable equilibrium phase. Further annealing results in the formation of a balanced crystalline $Ni_{1-x}Ti_x$ phase. Multilayer analysis by TEM did not observe complete amorphization. An amorphous phase formed in all samples, and then the reaction apparently stopped. The low-growth model predicts that for diffusion-limited growth, due to interface-limited growth, the thickness of the growth layer will change as the square root of time dynamics follows the linear time law. Long-term deviations may be caused by mechanical stress releases that affect the average concentration.

9.6 APPLICATIONS OF SHAPE MEMORY ALLOYS

Bulk and thin-film NiTi alloys have reached various levels of commercial development. For a wide range of critical applications, such as assembly of microsystems, endoscopes for microsurgery, and micromanipulators for injection of cellular drugs,

NiTi Thin-Film Shape Memory Alloys

the manipulation of microobjects requires high accuracy. They are the most demanding requirements in the field of MEMS. NiTi thin films have occupied important positions in microactuators and micromanipulators, couplings and fasteners, medical applications, adaptive materials and hybrid composites, and applications based on the high attenuation capabilities of SMA.

9.6.1 MICROVALVES AND MICROPUMPS

MEMS-based microvalves and pumps are attractive for many applications, such as implantable drug delivery, chemical analysis, and analytical instruments. TiNi membrane–based micropumps or microvalves have different designs, and most use TiNi membranes (diaphragms, microbubbles, etc.). Figure 9.11 shows an electric miniature valve. Here, SMA wires are connected to the gate and cantilever. In the activated state, the SMA wire is heated and shrunk by resistance, causing the cantilever to deflect and open the valve. In the closed state, the air flow is partially blocked by the door (Figure 9.11b). When power is removed from the SMA, the wire will cool, and the cantilever will stretch the SMA wire again, closing the port (Figure 9.11a). Figure 9.12 shows an electromagnetically driven vibrating diaphragm micropump using a diaphragm-type passive check valve. The electromagnetic operating mechanism consists of a permanent magnet attached to a diaphragm. When alternating current passes through the coil, Lorentz force moves the magnet up and down, causing the diaphragm to vibrate. The diaphragm valve consists of two valve parts

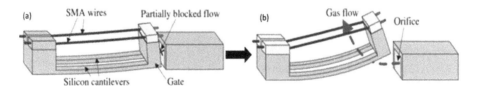

FIGURE 9.11 Illustration of an SMA wire microvalve in the closed state (a) and the open state (b). SMA, shape memory alloy.

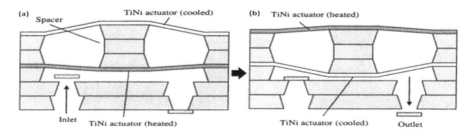

FIGURE 9.12 Operation and structure of an electromagnetically actuated micropump: (a) macroopening in cold condition (suction mode) and (b) microopening in heating condition (pumping mode).

with a thin diaphragm in the middle. Each valve consists of a valve seat and is injection molded. When the diaphragm is operated outward, the valve diaphragm moves upward from the seat of the inlet valve and fluid flows into the chamber (Figure 9.12a). As the diaphragm moves backward, the valve diaphragm returns to cover the inlet valve seat and moves away from the outlet valve seat, allowing fluid to flow out of the pump chamber (Figure 9.12b).

9.6.2 Microgripper and Microtweezer

In a variety of microsystem applications (such as microsystem assembly, microsurgical endoscopes, and cellular drug injection), it is essential to maintain very small objects with high accuracy. NiTi films are effective for these applications. The miniature fixture shown in Figure 9.13 consists of two relatively offset NiTi shape memory microactuators. One microactuator is trained as a bending actuator, whereas the other is a folded beam structure with linear actuation. The linear actuator has a deflection relative to the electric heating on the curved actuator, causing the jaws to open (Figure 9.13a). The heating of the bending actuator relative to the linear actuator also deflects, which causes the gripper to close (Figure 9.13b). Microfixtures with a thickness of 100 μm provide a maximum clamping force of 15 mN and a maximum displacement of 200 μm [31].

9.6.3 Biomedical Equipment

Stents play an important role in the medical industry. The stent can prevent the blockage of the passage in the living body. The typical operation of the stent is shown in Figure 9.14. Two NiTi thin film valve designs were developed, as shown in Figure 9.15. In smaller catheters, NiLi SMA films are used as its valve material.

9.7 ADVANTAGES AND LIMITATIONS OF NiTi THIN FILM

The main benefits of TiNi thin film include the following:

1. Higher power density
2. Larger displacement and actuation force

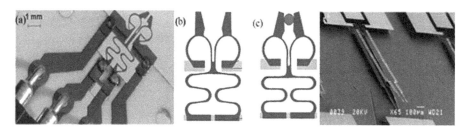

FIGURE 9.13 (a) NiTi microgripping instrument, (b) microgripper in open condition, (c) microgripper in holding state, and (d) free-standing TiNi-based film microtweezer structure that has both horizontal and vertical movements due to both shape memory and thermal effects [32].

NiTi Thin-Film Shape Memory Alloys

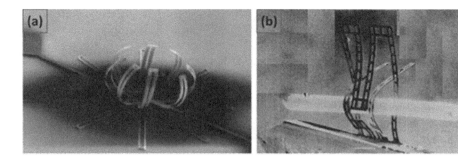

FIGURE 9.14 A typical procedure of stent inside the body of a human.

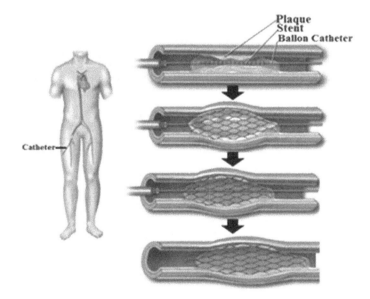

FIGURE 9.15 NiTi thin film heart valve designs (a) with sufficient coaptation height, (b) 7000TFX Perimount Magna 25 mm bovine mitral from Edwards Lifescience [33].

3. Lower operation voltage
4. Facilitating simplification of mechanisms with flexibility of design and creation of clean, friction free, and nonvibration movement
5. In vacuum deposition, reducing the particle density so that the mean free path for collision is long
6. In vacuum deposition, reducing the contaminants
7. In vacuum deposition, low-pressure plasma environment
8. Here, the use of magnets increases the percentage of electrons participating in the ionization event, increases the probability that the electrons will

contact argon, and increases the length of the electron path, thereby greatly improving the ionization efficiency

9. By using a magnet, it can reduce the voltage required to hit the plasma, adjust the uniformity, reduce the waffle heating caused by electron bombardment, and increase the deposition rate

10. Impurity content and smooth control of deposition rate can also produce thin films with different morphologies and crystal structures

Even with shortcomings such as low energy efficiency and low dynamic response, NiTi thin films are still considered to be the core technology for operating most MEMS devices in which high power and range are critical and under low load cycling conditions or intermittent action, and under extreme conditions, such as radioactive, space, biological, and corrosive conditions.

9.8 CONCLUSIONS

Shape memory materials continue to receive attention due to their great potential for active and passive applications in small- and large-scale industries. This chapter provides a comprehensive introduction to the technical aspects of alloys with shape memory and the inherent functions required of intelligent systems. A brief description of the historical background of SMAs has been explained. Compared with other modes of operation (i.e., electrostatic, thermal and piezoelectric), alloys with shape memory have the highest operating forces. The unique characteristics of NiTi alloys focus on in-depth research on all possible applications. NiTi films formed by sputtering have attracted great interest as powerful actuators in MEMS (such as microvalves, microfluidic pumps, and microgrippers). Phase diagrams are used to understand heat treatment procedures and to understand the relationship between phase transition temperature and microstructure, composition, and sediment development. Both the stress–strain temperature diagram and the NiTi phase diagram contribute to its possible application under different manufacturing and construction conditions. The crystallographic changes during the phase change during its surgery have been extensively studied. There are different types of technologies for NiTi film production. The physical properties of NiTi films can be clearly described using FESEM micrographs, GIXRD analysis, TEM micrographs, and AFM imaging. Nanoindentation and nanoscratch tests give the concepts of nanohardness, reduced modulus, elastic recovery during release, and abrasion. This chapter includes a wide range of application possibilities to realize the possibilities of future functional materials. Even with some flaws, NiTi films are still considered to be the core technology used to operate certain MEMS devices, and this has been clearly discussed.

REFERENCES

1. K. Otsuka, C. M. Wayman (1999) *Shape Memory Materials*, Cambridge University Press, New York 10011–4211, ISBN 0-521-66384-9.
2. K. Otsuka, X. Ren (1999) Recent developments in the research of shape-memory alloys, *Intermetallics*, **7**:511–528. DOI: 10.1016/S0966-9795(98)00070-3.

NiTi Thin-Film Shape Memory Alloys

3. K. Otsuka X. Ren (2005) Physical metallurgy of Ti-Ni-based shape memory alloys. *Progress in Materials Science*, **50**:511–678. DOI: 10.1016/j.pmatsci.2004.10.001.

4. O. K. Rediniotisa, D. C. Lagoudasa, R. Mania (2002) *Active Skin for Turbulent Drag Reduction*, Aerospace Engineering Department, Texas A&M University, College Station, TX 77843–3141, http://smart.tamu.edu/publications/docs/proceedings/2002/spie-2002-active-skin.pdf.

5. G. B. Kauffman, I. Mayo (1996) The story of nitinol: the serendipitous discovery of the memory metal and its applications. *The Chemical Educator*, **2**(2). ISSN 1430–4171, Springer-Verlag New York, Inc. DOI: 10.1007/s00897970111a.

6. S. Miyazaki, K. Otsuk (1989) Development of shape memory alloys. *ISIJ International*, **29**(5):353–377. DOI: 10.2355/isijinternational.29.353.

7. M. Nishida, C. M. Wayman, R. Kainuma, T. Honma (1986) Further electron microscopy studies of the Ti11Ni14 phase in an aged Ti-52at%Ni shape memory alloy. *Scripta Metallurgica*, **20**(6):899–904. DOI: 10.1016/0036-9748(86)90463-1.

8. R. Raghavendra Adharapurapu (2007) PhD Thesis: Phase transformations in nickel-rich nickel-titanium alloys: influence of strain-rate, temperature, thermomechanical treatment and nickel composition on the shape memory and superelastic characteristics, University of California, San Diego, http://escholarship.org/uc/item/7dt6n9p8.

9. A. Taylor, R. W. Floyd (1950) Precision measuremets of lattice parameters of non-cubic crystals. *Acta Crystallographica*, **3**(4):285–289. DOI: 10.1107/S0365110X50000732.

10. M. Nishida, C. M. Wayman (1987) Phase transformation in Ti_2Ni_3 precipitates formed in aged Ti-52 at. pct Ni. *Metallurgical Transactions A*, **18A**:785–799.

11. T. Hara, T. Ohba, K. Otsuka, M. Nishida (1997) Phase transformation and crystal structures of Ti_2Ni_3 precipitates in Ti-Ni alloys. *Materials Transactions, JIM*, **38**(4):277–284. DOI: 10.2320/matertrans1989.38.277.

12. S. Miyazaki, K. Otsuka (1986) Deformation and transition behavior associated with the R-phase in titanium-nickel alloys. *Metallurgical Transactions A: Physical Metallurgy and Materials Science*, **17**(1):53–63. DOI: 10.1007/BF02644442.

13. T. Saburi, S. Nenno, T. Fukuda (1986) Crystal structure and morphology of the metastable X-phase in shape memory alloys. *Journal of Less-Common Metals*, **125**:157–166. DOI: 10.1016/0022-5088(86)90090-1.

14. T. Tadaki, Y. Nakata, K. Shimizu, K. Otsuka (1986) Crystal structure, composition and morphology of a precipitate in an aged titanium-51 at.% nickel shape-memory alloy. *Transactions of the Japan Institute of Metals*, **27**(10):731–740. DOI: 10.2320/matertrans1960.27.731.

15. W. Tirry, D. Schryvers, K. Jorissen, D. Lamoen (2006) Electron-diffraction structure refinement of Ni 4 Ti 3 precipitates in Ni 52 Ti 48. *Acta Crystallographica Section B Structural Science*, **438**:517. DOI: 10.1107/S0108768106036457/lc5049sup2.txt.

16. X. Wang (2007) Crystallization and Martensitic Transformation Behavior of NiTi Shape Memory Alloy Thin Films, Ph.D. dissertation The School of Engineering and Applied Sciences, Harvard University Cambridge, Massachusetts.

17. A. M. Loccia, R. Orru, G. Cao, Z. A. Munir (2003) Field-activated pressure-assisted synthesis of NiTi. *Intermetallics*, **11**:555–571. DOI: 10.1016/S0966-9795(03)00043-8.

18. S. N. Kulkov, Y. P. Mironov (1995) Martensitic transformation in NiTi investigated by synchrotron X-ray diffraction. *Nuclear Instruments and Methods in Physics Research A*, **359**:165–169. DOI: 10.1016/0168-9002(94)01644-5.

19. S. Miyazaki, K. Nomura, A. Ishida (1995) Shape memory effects associated with the martensitic and R-phase transformations in sputter-deposited Ti-Ni thin films. *Journal de Physique IV*, **5**(C8):C8-677–C8-682. DOI: 10.1051/jp4/199558677.

20. X. B. Wang, B. Verlinden, J. V. Humbeeck (2014) R-phase transformation in NiTi alloys. *Materials Science and Technology*, **30**(13a):1517–1529. DOI: 10.1179/1743284714Y. 0000000590.

21. M. Tomozawa, H. Y. Kim, S. Miyazaki (2006) Microactuators using r-phase transformation of sputter-deposited Ti-47.3Ni shape memory alloy thin films. *Journal of Intelligent Material Systems and Structures*, **17**:1049–1058. DOI: 10.1177/1045389X06064883.

22. J. Uchil, K. K. Mahesh, K. G. Kumar (2002) Electrical resistivity and strain recovery studies on the effect of thermal cycling under constant stress on R-phase in NiTi shape memory alloy. *Physica B*, **324**:419–428. DOI: 10.1016/S0921-4526(02)01462-X.

23. G. Eggeler, J. Khalil-Allafi, S. Gollerthan, C. Somsen, W. Schmahl, D. Sheptyakov (2005) On the effect of aging on martensitic transformations in Ni-rich NiTi shape memory alloys. *Smart Materials and Structures*, **14**:S186–S191. DOI:10.1088/0964–1726/14/5/002.

24. C. P. Frick, T. W. Lang, K. Spark, K. Gall (2006) Stress-induced martensitic transformations and shape memory at nanometer scales. *Acta Materialia*, **54**(8):2223–2234. DOI: 10.1016/j.actamat.2006.01.030.

25. L. Zhang, C. Xie, J. Wu (2007) Oxidation behavior of sputter-deposited Ti-Ni thin films at elevated temperatures. *Materials Characterization*, **58**:471–478. DOI: 10.1016/j.matchar.2006.06.011.

26. J. E. Jongste, M. A. Hollanders, B. J. Thijsse, E. J. Mittemeijer (1988) Solid state amorphization in Ni/Ti multilayers. *Materials Science and Engineering*, **97**:101–104. DOI: 10.1016/0025-5416(88)90020-1.

27. S. H. Whang, L. T. Kabacoff, D. E. Polk, B. C. Giessen (1980) A technique for the measurements of young's modules of small metallic glass samples. *Journal of Materials Science*, **15**:247. DOI: 10.1007/BF02811717.

28. W. J. Meng, B. Fultz, E. Ma, W. L. Johnson (1987) Solid state interdiffusion reactions in Ni-Ti and Ni-Zr multilayered thin films. *Applied Physics Letters*, **51**:661. DOI: 10.1063/1.98326.

29. M. A. Rahman, T. Osipowicz, D. Z. Chi, W. D. Wang (2005) Observation of a new kinetics to form Ni3Si2 and Ni31Si12 silicides at low temperature (200°C). *Journal of the Electrochemical Society*, **152**:G900–G902. DOI: 10.1149/1.2077329.

30. K. R. Loger, L. D. Miranda, A. Engel, M. Marczynski-Buhlow, G. Lutter, E. Quandt (2014) Fabrication and evaluation of nitinol thin film heart valves. *Cardiovascular Engineering and Technology*, **5**(4):308–316. DOI: 10.1007/s13239-014-0194-6.

31. S. S. Lyer, Y. M. Haddad (1994) Intelligent materials–an overview. *International Journal of Pressure Vessels and Piping*, **58**:335–344. DOI: 10.1016/0308-0161(94)90070-1.

32. W. M. Ostachowicz, M. P. Cartmell, A. J. Zak (2001) Statics and dynamics of composite structures with embedded shape memory alloys, In: *Proceedings of the International Conference on Structural Control and Health Monitoring*, SMART 2001, Warsaw.

33. Y. Fu, H. Du, W. Huang, S. Zhang, M. Hu (2004) TiNi-based thin films in MEMS applications: a review. *Sensors and Actuators A*, **112**:395–408.

10 Carbon Fibers
Surface Modification Strategies and Biomedical Applications

Suneev Anil Bansal
MAIT, Maharaja Agrasen University

Javad Karimi
Shiraz University

Amrinder Pal Singh and Suresh Kumar
UIET, Panjab University

CONTENTS

10.1 Surface Treatment 207
    10.1.1 Surface Oxidation 208
    10.1.2 Surface Coating 210
10.2 Applications of Carbon Fibers 212
    10.2.1 Carbon Fibers and Composite Implant for Bone 213
    10.2.2 Carbon Fiber in Tissue Growth 213
    10.2.3 Dental Implants 214
    10.2.4 Regenerative Medicine 214
    10.2.5 Carbon Fibers in Drug Delivery 215
    10.2.6 Carbon Fibers in Biomedical Sensor 216
    10.2.7 Carbon Fibers Composites 217
10.3 Summary 219
References 220

10.1 SURFACE TREATMENT

A suitable surface treatment process can greatly enhance the mechanical properties of carbon fibers (CFs) [1]. Most of the surface treatment processes are still confidential due to high commercialization. Suitable surface treatment methods can enhance the suitability of CFs for the final application. Figure 10.1 shows the detailed options available for surface treatments.

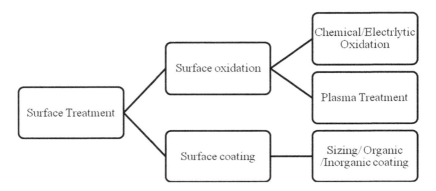

FIGURE 10.1 CFs surface treatment methods. CF, carbon fiber.

10.1.1 Surface Oxidation

The most commonly used surface treatment processes for CFs are gaseous oxidation and liquid oxidation treatments [2]. The liquid treatments can enhance the shear strengths of CFs, up to twofold, with little reductions (4%–6%) in CFs' tensile strengths [3].

Due to its inexpensive nature, fast and efficient liquid oxidation treatment methods are being widely used. Figure 10.2 shows the treatment process showing the use of several washing baths. Using Faraday's law, it can be estimated that 96,500 C will release 1 g equivalent of O_2. The line speed influences the duration of treatment expressed in C/m. Ammonium sulfate is usually used as an electrolyte in the commercial surface treatment of CFs. The presence of carbonyl-containing groups (like –COOH) helps to form a smooth CF surface. The –COOH group also improves the cohesion of the fiber with the resin used in the composite. Warm washing is performed to remove any extra electrolytes present in the CFs. The CFs are then passed to the next process, that is, treatment with water at a constant rate of water flow. The demand for superior surface treatment of CFs has significantly increased with an increase in demand for high-quality CFs and their composites.

Most of the CFs are prepared by polymerizing natural or synthetic materials. These solutions are then sized into fibers. The prepared CFs are having the very low specific surface area and adhesion properties.

Due to the requirements, sometimes it is not possible to modify the surface of fiber for the final application. In those cases, a thin film of some other materials is applied on the surface of the fiber to make it chemically active for further processing. From an industrial point of view, electrolytic oxidation is the most widely used oxidation process. An electrolytic solution such as HNO_3 is used to etch the surface of CFs which activates the surface of CFs and helps to attach functional groups on the surface of CFs. The attachment of functional groups on the surface of CFs can be controlled both qualitatively and quantitatively [5]. This is highly scalable and can be integrated well with the existing CFs manufacturing process. Electrolytic oxidation produces a lot of waste solution, which conflicts with environmental issues restricting its mass production.

Carbon Fibers

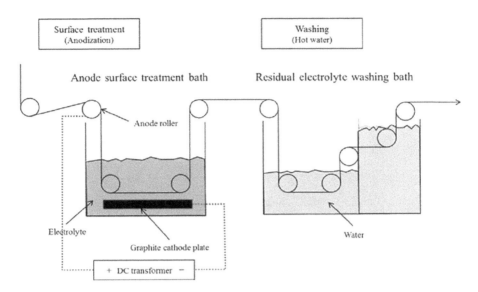

FIGURE 10.2 Surface treatment process for carbon fibers [4].

Plasma treatment is a cleaner method to activate the surface of CFs. Oxygen, argon, ammonia, and so on treated with plasma ions are used to activate the surface of CFs. Surface modifications depend on the type of plasma and type of CFs to be treated. Figure 10.3 shows the effect of plasma treatment on CFs. After plasma treatment, CO and COOR intensity have increased. CO is actually C=O double bond. This shows that plasma treatment increased the oxidation of CFs. The intensity of the C–O peak is decreased too. The plasma treatment is performed in the present case transferred C–O single bond to C=O double bond. Using the same method, hydroxyl and carboxyl groups can be obtained while working with polyacrylonitrile (PAN)-based CFs. In pitch-based system, only hydroxyl groups can be obtained. Plasma treatment successfully and precisely controls the surface properties. The type of plasma used and prevailing conditions, however, influence the final outcome.

Methods stated earlier for surface treatment of CFs need to be applied carefully. Unusual long exposure of oxidation methods may lead to loss of weight, erosion, and quality degradation of CFs produced. Another treatment in this field, although less used, is radiation treatment. There are a variety of radiation treatments available to treat the surface of CFs. Radiations modify the behavior of CFs either by electronic excitation or by the displacement of molecules and atoms. These effects change the crystal structure of the final CF [7]. Few of the electrons may leave the surface of CFs excited by radiation therapy. This excitation creates the activated sites that finally help to attach the polymer. Researchers successfully used gamma radiation for surface treatment of PAN-based materials [8]. To understand the oxidation using gamma ray method, researchers compared the results of gamma ray oxidation with air oxidation. Composites prepared with gamma-treated CFs and air-treated CFs were tested for mechanical characterization. It is observed that gamma-treated CFs composite was having better CFs–polymer bonding. The likely cause of output

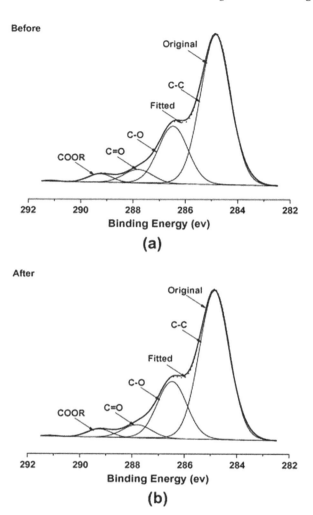

FIGURE 10.3 XPS spectrum of (a) untreated and (b) plasma-treated CF [6]. CF, carbon fiber; XPS, X-ray photoelectron spectroscopy.

was attributed to the presence of more carboxyl group in gamma-treated CFs. The ion beam can also be used to modify the surface of CFs (Figure 10.4) [9]. Ar+ ion is used as a radiation medium to modify the surface of CFs used for epoxy matrix–based composite material. Fourier transform infrared characterization showed a new carboxyl peak due to the formation of a new functional group. The new and treated functionalities result in synthesis of a stronger epoxy-based composite materials.

10.1.2 Surface Coating

Sizing of CFs is a process by which the desired surface finish or required size of the fiber can be achieved. Mostly coating of a polymer is done over the CFs to get

Carbon Fibers 211

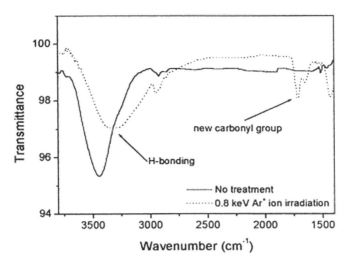

FIGURE 10.4 FTIR of Ar+ ion-treated CFs [9]. CF, carbon fiber; FTIR, Fourier transform infrared.

the required results [10,11]. The coating is done on the surface primarily for the protection of surface, wettability, and alignment. Zeng et al. sized the CFs with different agents to produce superior composites [12]. Emulsifier agent, with 10, 15, and 20 wt.% addition for sizing, gives encouraging results. The agents are also used with nanoparticles for sizing [13]. Carbon nanotubes (CNTs) are used to fabricate CNT–CF hybrids to be used as fillers [14]. Chemically modified CNTs, with added functionalities, are used. Nonemulsion are also used to size CFs [15]. A polyimide and epoxy, a combination of thermoplastic and thermosetting plastic, are also used to size CFs [16]. This sizing has produced a heat-resistant coating. The enhancement in wear strength is almost doubled by the sizing treatment.

Coating with inorganic material is also very popular for CFs. The most common coating used is metal oxide coating. The coating is done by condensation and hydrolysis reaction of alkoxides of metals. Popular methods used for coating are immobilization of nanoparticles and very thin-layered coating. However, it is known that each process has its demerits and merits. The effect of metal coating on CFs produces far better results than metal oxide and bare CFs. The coating enables the interface of CFs and matrix material to create a strong bond. The covalent bonds on the surface of CFs can be stabilized by sol–gel method. The advantages of using the sol–gel methods are numerous such as simple, recyclable, complex shape substrate, and coating up to temperature 100°C [17,18]. Functionalization of CFs can also be achieved by immobilizing nanoparticles. This technique is used by controlling the charge over the surface of CFs. Solution pH can also be used to tune the properties of CFs as the nanoparticles are not bonded with surface but attached to the surface due to charge on the surface of CFs. Shiba et al. [19] have been able to synthesize titanium-based nanoparticles. These nanoparticles are used to control light scattering from CFs. This property is used to develop light-based optical materials.

There are a variety of coatings available to produce the biocompatibility of CFs. The coating of hydroxyapatite (HAp) nanoparticles are used to enhance the biocompatibility of CFs. The primary use of this coating is in the field of implants. HAp is widely used for drug delivery, bone filling, and so on [20,21]. Studies are also done to investigate the protein absorption state on the surface using HAp nanoparticles. These studies identified HAp, as a capable material, to be used on the surface of CFs. The resulting material becomes biocompatible.

Fluorination is another technique to be used to enhance the surface of CFs. The degree of fluorination depends upon the type of CFs used and the type of process used [22]. Well-graphitized structure or higher temperature and pressure can be used to obtain a high amount of fluorine in intercalated form. At low temperature, the fluorine makes C–F bond on the surface of CFs [23]. Further addition of fluorine stops due to this bonding. The effect of degree of fluorination on properties of CFs was also studied [22]. Different methods have been used to add fluorine on the surface of CFs. The reaction has been performed in an electric furnace in a reactor made of nickel. Fluorine and oxygen gas have been introduced after evacuation. The pressure has been set to 0.1 MPa, and time to reaction was 2 h. Studies have shown that there are considerable changes at the surface of CFs. Even mild fluorine gas has been able to modify the surface of CFs.

10.2 APPLICATIONS OF CARBON FIBERS

CFs is a generic term related to a family of fibers synthesized using the pyrolysis of mainly organic precursor fibers such as rayon, PAN, and pitch in an inert environment. CFs have high crystallinity and strong covalent bonds with a graphitic structure. CFs have highly anisotropic mechanical properties. The mechanical properties are found to be better along the axis (longer axis) than that perpendicular to the axis. The layers along perpendicular direction are attached by van der Waals forces, which are considered as weak forces. To achieve high mechanical properties such as high modulus in CFs, the graphitic crystal orientation requires improvement. This improvement in orientation can be achieved by thermal treatments. Stretching treatments contribute to the orientation improvement of graphite in the fibers. Low density and high mechanical properties of CFs help to achieve high strength and high modulus. The values as shown in Table 10.1 are even better than that of materials such as steel. High strength and high modulus of CFs make them a suitable candidate for applications such as high-performance composites used in aerospace and aircraft industries. The elastic recovery of CFs is great and even can achieve full recovery. CFs also have wonderful fatigue resistance. CFs, even at room temperature, have very good moisture resistance and very good chemical resistance. CFs on the other side bear a drawback of oxidization at higher temperatures (350°C–450°C). In addition to the previously discussed mechanical applications, CFs are suitably used in biocompatible applications. These applications are proven clinically and by experimental trials [24].

Biologically studied and clinically proved CFs find a wide variety of applications. Volumetric recognition of biological molecules can be carried out by electrically conductive CFs. CFs are also used for neural recording [27,28]. Hard tissues

Carbon Fibers

TABLE 10.1
Mechanical Properties of CF, High-Modulus CF, and Steel [25,26]

Type	Gravity (g/cm³)	Tensile Strength (GPa)	4 Modulus (GPa)
CF	1.6–2.2	1.5–5.65	228
High-modulus CF	2–3	3.4–6.4	345
Steel wire	7.9	2.39	210

CF, carbon fiber.

and soft tissues can be healed by CFs too [25]. Ligament repair has also been performed by the use of CFs. Replacement of ligament is also tested using CFs. The same technique is also used for knee reconstruction. Although CFs are mechanically sound along axis, still there is a drawback in shear strength that may lead to debris and fragments. That is why CFs have not been accepted by the Food and Drug Administration for internal ligament replacement. CFs-based composites are also used for bone fracture repair [29].

10.2.1 CARBON FIBERS AND COMPOSITE IMPLANT FOR BONE

CFs and their composites have been widely studied for application as bone implants [30]. CFs provide requisite strength to the bone. CFs possess high strength and have a density comparable with that of bone. These materials have been developed with epoxy-based composite materials. Earlier the titanium-based implants have been used for the same purpose. CFs-based implants have been compared with titanium-based implants. Unidirectional composite rods of epoxy–CF having 1.5 mm diameter are placed for 2 weeks using a rat tibia test [31]. Unidirectional composite with 60% fiber and 40% epoxy has also been manufactured and studied.

10.2.2 CARBON FIBER IN TISSUE GROWTH

Repair or replacement of ligaments and tendons is one of the earliest medical uses of CFs. CFs do not hinder tissue growth and can be used in biomedical engineering applications. However, the mechanism of removal and disintegration of CF implants is still questionable. Few reports have shown that CFs can break into smaller fragments and move into the nearest lymph without any detrimental effect. CFs in all instances assisted the regeneration of tendon and ligament. Although Morris et al. [32] have reported that fragmentation of fibers hardly occurred, debris of implant was not found at nodes. CFs can assist the tissue growth. CFs-based implants have been tested in vivo and showed good results. However, few researchers have claimed prosthetic materials to be superior to CFs. Blazewicz et al. [33] have evaluated the synthesis, structure, and property relationships of CFs in connection with the tissue response. The study shows that material, orderly arrangement of crystal, microstructure property, and surface conditions influence the tissue response. Higher crystallinity and organized structure of graphite in CFs assimilate with more difficulty in

214 Advanced Manufacturing and Processing Technology

the body, and tiny particles, from these materials, are found in the regional nodes. Smaller crystallite size (low carbonized) CFs have experienced low fragmentation and reacted well in the biological environment even subsequently absorbed in the plantation site. Surface of CFs is capable of supporting phagocytosis. Process is assisted by macrophages and enhanced by acidic groups. The nature of CF surface also influences the healing rate of the wound on the bone.

10.2.3 DENTAL IMPLANTS

The use of titanium-based materials for dental implants can result in complications such as titanium hypersensitivity. Schwitalla et al. [34] have reviewed the materials used in dental applications. The reported literature has analyzed the CF-reinforced polyether ether ketone (PEEK) (CFs–PEEK) dental implants stress distribution by the three-dimensional finite element model. It has demonstrated higher stress because of reduced stiffness in comparison with titanium. Comparison of CFs–PEEK with titanium-coated CFs–PEEK implants has been evaluated for 4–8 weeks planted in femurs. The TiC (titanium-coated) implants have shown higher bone–implant contact (BIC) rates; therefore, CFs have potential applications in dental care.

10.2.4 REGENERATIVE MEDICINE

Regenerative medicine is an important area of research these days. Various reports are published on the use of regenerative medicine in the field of cell therapy, gene therapy, cytokine therapy, and so on [35,36]. The approach used in these studies is to hold the cells in a scaffold at a local site that facilitates new tissues for regenerative growth. The development of this scaffold is very important. CFs find important roles in this field as these are very strong, do not corrode like metals or other materials, and are bioinert. CFs are also tested for positive bioactivity, as bone tissues can be regenerated around CFs [37]. Studies of in vitro showed that the use of CFs activates tissue growth and improves adhesion [38]. Possible bioactivities of CFs in other parts of the body are also under investigation.

Using CFs for tendon repair has been started with animal studies in the year 1980. Studies performed on tendon tissue of animals showed the good repair. Studies on horse have shown a good tendon repair using CFs. Studies have started in later years to repair tendon using CFs. Absorbable classes of materials/composites have also been studied. The bottleneck to use this mechanism for tendon repair is the detachment of CFs, and it is scattering to nearby areas on prolonged usage. Further studies of CFs for ligament repair have shown good biocompatibility. The suitability of CFs to repair anterior cruciate ligament is questionable. The reason for this may be due to its very less repair capability and nature of CFs. Mäkisalo et al. have studied the connective tissue reaction of CFs and polypropylene on 30 rats sample [39]. Materials have been implanted at femoral bone tunnel and skin. There has been prominent granulation tissue around CFs. In the year 1989, a study was published to repair bone injuries and tissue regeneration on the animal models [40]. The study was performed to repair articulate cartilage. In 1999, cartilage

Carbon Fibers

defects were embedded with CFs, and performance was checked after 1 year. The results were encouraging [41]. However, in this repair, the material of the scaffold is very important, as the activity of cartilage to regenerate is very low. Spinal cage fixation was also done using CFs. The abdominal wall was also successfully repaired using CFs. A new method was employed to fabricate carbon-based fine implants [42]. The primary focus of the method was to promote tissue growth. The material was implanted in a calf. In vivo historical data were captured. The results were encouraging. Nanosized carbon materials have also been reported to be used to prepare a thin CF web to be used as a scaffold [43]. CFs help cells to attach on the surface of fibers. Blood vessels can easily penetrate from this thin CFs. Their ability to retain their shape is very high, and control of shape tissue regeneration is good [44]. Thin CFs web was synthesized by electrospinning using PAN as a precursor. There was a thermal treatment in argon at 1000°C. Toxicity of these materials is low as there is low content of metals. These materials are also suitable to synthesize cylindrical implants. Thin CFs web efficacy with bone tissue generation was also studied. All implants were placed at mice's dorsal muscles. After 2 weeks of implant, a small formation of bone was observed. The comparative study between collagen groups was also performed. In 1 week's study, there was no difference in the regeneration of bones. Both types of materials showed inflammatory reactions and foreign body reactions. At the completion of 3 weeks, normal bone tissue of mature bone developed. Thin CFs web material showed thicker bone trabeculae than collagen materials. At 3 weeks, studies of BMC (bone mineral content) and BMD (bone mineral density) showed that thin CFs web was more appropriate material for the scaffold.

10.2.5 Carbon Fibers in Drug Delivery

Drug delivery is a very attractive method in medicine that refers to the transfer of a medicinal compound in the existing body to induce a desired therapeutic effect with the least side effects. Drug connection to the carrier, transmission drug carrier in the body, and the precise, controlled, and targeted drug delivery, for example, in cancerous tissues are very important [45]. CFs, carbon nanotubes, and other nano- or microcarbon structures have good mechanical properties such as quite great surface area, high tensile strength, high electrical conductivity, high thermal conductivity, and respectable flexibility, which are very useful to drug delivery (Figure 10.5) [46].

The kind of matter is highly important on the surface of the CF in the delivery system, and it is related to the type of matter and the contact angle. In 2017, Tangboriboon et al. have found that potential of obtained carbon drug delivery system in the water is excellent in the vitamin C, and it was low about water oil (Figure 10.6). The interaction angle values of water, vitamin C, and oil were 8.3, 23.5, and 36.3 degrees, respectively [47].

Tangboriboon showed in another experiment that drug delivery system with addition of CF to latex showed high electrical conductivity and good tensile strength. It means that a drug can be rapidly transmitted within the target cells that are stimulated by the electrical current 3.

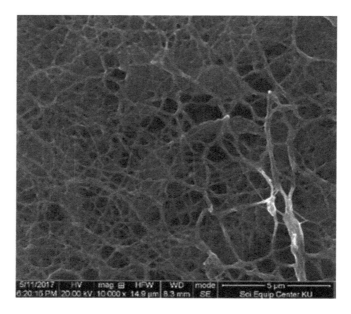

FIGURE 10.5 Scanning electron microscopy of carbon drug delivery prepared from eggshell.

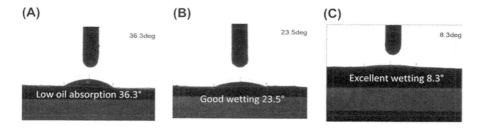

FIGURE 10.6 Absorption and contact angle (A: oil, B: vitamin C, and C: water) on drug delivery.

10.2.6 CARBON FIBERS IN BIOMEDICAL SENSOR

Medical sensors are important components of diagnostic medical equipment that are found in many medical centers, clinical labs, and hospitals. These advanced structures in clinical medicine and biological research are used to conduct monitoring on a wide range of vital physiological variables such as blood pressure, temperature, and blood glucose [48,49]. In another definition, medical sensors are tools that identify specific biological, chemical, and physical processes and report a tangible form after analysis of these data. Biomedical sensors help in monitoring of the safety of medicine by reviewing the effectiveness under controlled environmental conditions [50].

Carbon Fibers 217

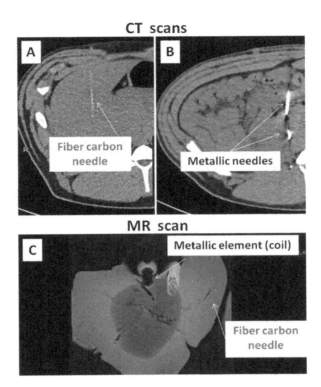

FIGURE 10.7 In vivo CT scans of pig liver with (A) the advanced carbon fiber needle (no artifact), and (B) communal needles for percutaneous measures (artifact). (C) Assessment between the noticeable artifact affected by a metallic component and the image of the fiber carbon needle (no artifact).

Saccomandi and coworkers in 2017 showed application of CF at temperature monitoring in a specific texture and during certain time. They too showed that CF needle without any artefact on the images has been tested in both MR and CT scanner, differently than numerous other tools presently used for percutaneous techniques (Figure 10.7) [51].

McConville and coworker in 2018 designed an accurate diagnosis of peroxide using surface modification of CF. Natural CF, while conductive, can, however, be a moderately poor sensor with no intrinsic discrimination and require alteration with chemical portions or enzymes. Treatment of highly operative Pd covering on CF offers a facile style to sensor design [52].

10.2.7 Carbon Fibers Composites

A combination of carbon materials with CFs produces a composite that is called CFs/carbon composite. The reinforcement material in these composites is CF. Although it is used in industrial applications such as aircrafts, brake, rockets, and so on, its properties are also favorable for bioapplications due to its biocompatibility, durability, and chemical stability. In vivo and in vitro studies of CFs showed CFs/carbon composites possess biocompatibility [53–55]. CFs with metal coating have also been used for bioapplications. CF with cobalt–nickel–chrome alloy showed good biocompatibility [56]. The study was able to successfully develop biocompatible hybrid cerebral aneurysm clip. Tantalum-based CFs composite also showed good bone growth. Using hydroxyapatite with CFs was also used to develop bone grafts and tested good for bioapplications. Other materials with CFs in the different domains were also studied and found good for their biocompatibility.

Composites of CFs with PEEK in bioapplications are also widely researched [57,58]. In the year 1990, PEEK with CFs was used in a spinal cage to improve the strength. The clinical trial for the same was also performed. The outcome of these studies helped to use PEEK–CFs composite to be used for spinal fixation [59]. Pure PEEK materials bear a disadvantage of insufficient strength. To enhance the strength of PEEK materials, the CFs reinforcement was done. Other materials used for this kind of composite synthesis are titanium, stainless steel, and so on. But PEEK/CFs have an advantage of similar specific modulus to the bone. Due to that, it does not produce interface disruption while usage. It also reduces the risk of bone loss. In this regard, various research groups presented studies to compare metallic implants with nonmetallic implants. Rohner et al. in 2005 [60] studied the performance of CFs–PEEK in vivo with titanium-based implants. The CFs–PEEK composite was used at a fixed angle as a high stiffness snake plate. Comparable titanium-based plate was used as a seven-hole locking compression plate. The sheep trials were done in two groups. Radiology was used to measure callus dimension at 2, 4, 6, and 8 weeks. It was concluded from the study that PEEK-based materials were a very good alternative as compared with metal-based materials.

To use these materials, various sterilization methods were also proposed. Godara et al. [61] studied the influence of sterilization on PEEK–CFs composite. The polymer composite used for that study was PEEK reinforced with CFs. To understand the behavior, nanoindentation technique was used. Nanoscratching was also done. As these products are planted inside the body, it is very important to understand the sterilization process. Two types of sterilization processes using steam and gamma radiation were used. Greater emphasis was given to the interphase of matrix and CFs. There were hardly any significant changes after both the sterilization processes. Out of two, steam sterilization had a small effect on PEEK matrix at the interface regions. Even after a lot of research, PEEK–CFs-based composite are still under research for biomedical applications.

Development of polyethylene-based highly wear-resistant medical/artificial parts with CFs reinforcement started in the year 1980. Artificial hip joints and caps of knees were studied. There was a disadvantage of using CFs with polyethylene (PE). The adherence of CFs with PE was not very good. This led to the wear of PE, very rapidly resulting in early replacement of joints. After that, the usage of PE–CFs was stopped. Progress can be done by settling this issue. Investigation at that moment can be an area to improve these days. Experience of failure in the past can also be used to develop new products in this area.

Thermosetting plastic bears advantage of very high strength to weight ratio. As compared with other thermoplastic, the adhesion of thermosetting plastic with fiber

Carbon Fibers

is very high [62]. This adhesion results in very good mechanical performance during service life. Although this kind of processing requires high cost, its application is justified only if that high cost can produce that kind of application. Medical application is one area where this high cost can be justified. Epoxy, a thermosetting plastic cure with hardener, is used for these kinds of applications. Sell et al. [63] studied the CFs epoxy plates with CFs polysulfone rivets. CFs epoxy–based composites also hold the advantage over metal plates under cyclic loading. The advantage of having elasticity modulus close to the bone is also there. The study was also aimed at replacing the metal screws with CFs-based screws. It is very difficult to drill fine threads in the CFs–epoxy composite, which lead to failure. The study reported good results at low tensional loads but failed at high tensional loads. Recently, Bedheri et al. [64] studied CFs–epoxy composite for bone fracture applications. Group studied the bone gene formation and cell viability of epoxy reinforced with CFs plate in comparison with a metal plate (stainless steel). Stainless steel plates are commonly used for these types of applications. The cell viability of the CFs–epoxy–plate was as good as metal plate. No cytotoxicity was observed too. There was no adverse effect on gene expression levels. This study put Flax/CFs/epoxy-based plates as a future candidate for bone fracture repair. Osseointegration success from titanium-based materials is very promising. Peterson [65] reviewed osseointegration success of titanium-based materials and put forward a CFs–epoxy-based composite for the same type of applications. Presently, titanium is a very good biocompatible material for implants, joints, or outplant applications. TiO_2 produced at the surface layer of titanium can be attributed to its biocompatibility. However, there are the cases of surface corrosion in case of titanium that produces inflammation and loosening at the bone and generates infections. These problems can lead to implant failure. These problems can be rightly attacked by providing a coating on the surface or using non–metal-based implants such as epoxy. Epoxy–CFs can be better used in this area.

Other composites synthesized with CFs are listed in Table 10.2. CFs/PMMA (poly(methyl methacrylate)) finds its application in cups and joints of the body parts. Polystyrene with CFs has a variety of applications in the area of hips, nail, and screws for biomedical fixation, plates for bone fixation, joints, etc. Thermosetting polymer epoxy reinforced with CFs finds application in the field of dental implants, joints, and hip joints. Plates of bone are manufactured with PP (polypropylene), PE, nylon, PLA (polylactic acid), and PET (polyethylene terephthalate) reinforced with CFs. All these are either clinically proven or under progress for clinical trials.

10.3 SUMMARY

Over the past half-century, CFs of various types, form, and compatibility have been synthesized and used in many applications. Chemical treatment and surface treatment processes can help in preparing biocompatible fibers. Modern treatment processes enhance the use of CFs in advanced bioapplications. The development of CFs-based materials is quite relevant due to their high biocompatibility and close physical dimension to the bone. A lot of products manufactured from CFs are under trials so that they can be used for real application. CFs are used in a variety of forms in biomedical applications in pure form with various dimensions or in composite

TABLE 10.2

Various CFs-Based Composites and Their Biomedical Applications [66]

S. No.	CFs Composite	Biomedical Application
1	CFs/UHMWPE	Acetabular cups
		Articular surfaces in joint applications
2	CFs/PMMA	Bone plates
		Articular surfaces in joint applications
3	CFs/PS	Hip stems
		Intramedullary nails and screws
		Bone plates
		Articular surfaces in joint applications
		Lumbar interbody fusion
4	CFs/epoxy	Hip stems
		Dental Implants
		Articular surfaces in joint applications
		Lumbar interbody fusion
5	CFs/PP	Bone plates
6	CFs/PE	Bone plates
7	CFs/nylon	Bone plates
8	CFs/PBT	Bone plates
9	CFs/PEEK	Hip stems
		Intramedullary nails and screws
		Lumbar interbody fusion
		Bone plates
10	CFs/PLA	Bone plates
11	CF/carbon	Hip stems

CF, carbon fiber; PBT, polybutylene terephthalate; PE, polyethylene; PEEK, polyether ether ketone; PLA, polylactic acid; PMMA, poly(methyl methacrylate); PP, polypropylene; PS, polysulfone; UHMWPE, ultrahigh-molecular-weight polyethylene.

form with matrix materials. There are a lot of scopes in the field of biomedical engineering to use carbon-based fibers alone and in the combination with other materials.

REFERENCES

1. Yao X, Gao X, Jiang J, Xu C, Deng C, Wang J. Comparison of carbon nanotubes and graphene oxide coated carbon fiber for improving the interfacial properties of carbon fiber/epoxy composites. *Compos Part B Eng* 2018;132:170–7. doi:10.1016/j.compositesb.2017.09.012.

Carbon Fibers

2. Yao S-S, Jin F-L, Rhee KY, Hui D, Park S-J. Recent advances in carbon-fiber-reinforced thermoplastic composites: A review. *Compos Part B Eng* 2018;142:241–50. doi:10.1016/j.compositesb.2017.12.007.
3. Drzal LT, Rich MJ, Lloyd PF. Adhesion of graphite fibers to epoxy matrices: I. The role of fiber surface treatment. *J Adhes* 1983;16:1–30. doi:10.1080/00218468308074901.
4. Park S-J, Meng L-Y. *Surface Treatment and Sizing of Carbon Fibers*, Springer Netherlands; 2015, pp. 101–33. doi:10.1007/978-94-017-9478-7_4.
5. Fitzer E, Jäger H, Popovska N, Von Sturm F. Anodic oxidation of high modulus carbon fibres in sulphuric acid. *J Appl Electrochem* 1988;18:178–82. doi:10.1007/BF01009259.
6. Rhee KY, Park SJ, Hui D, Qiu Y. Effect of oxygen plasma-treated carbon fibers on the tribological behavior of oil-absorbed carbon/epoxy woven composites. *Compos Part B Eng* 2012;43:2395–9. doi:10.1016/j.compositesb.2011.11.046.
7. Tiwari S, Bijwe J, Panier S. Gamma radiation treatment of carbon fabric to improve the fiber–matrix adhesion and tribo-performance of composites. *Wear* 2011;271:2184–92. doi:10.1016/j.wear.2010.11.032.
8. Wan YZ, Wang YL, Huang Y, Luo HL, Chen GC, Yuan CD. Effect of surface treatment of carbon fibers with gamma-ray radiation on mechanical performance of their composites. *J Mater Sci* 2005;40:3355–9. doi:10.1007/s10853-005-2844-4.
9. Park S-J, Seo M-K, Rhee K-Y. Effect of Ar+ ion beam irradiation on the physicochemical characteristics of carbon fibers. *Carbon NY* 2003;41:592–4. doi:10.1016/S0008-6223(02)00395-0.
10. Vara H, Collazos-Castro JE. Enhanced spinal cord microstimulation using conducting polymer-coated carbon microfibers. *Acta Biomater* 2019;90:71–86.
11. Motlagh MS, Mottaghitalab V. The charge transport characterization of the polyaniline coated carbon fabric as a novel textile based counter electrode for flexible dye-sensitized solar cell. *Electrochim Acta* 2017;249:308–17.
12. Zhang RL, Huang YD, Liu L, Tang YR, Su D, Xu LW. Effect of emulsifier content of sizing agent on the surface of carbon fibres and interface of its composites. *Appl Surf Sci* 2011;257:3519–23. doi:10.1016/j.apsusc.2010.11.066.
13. Meguid S, Sun Y. On the tensile and shear strength of nano-reinforced composite interfaces. *Mater Des* 2004;25:289–96. doi:10.1016/j.matdes.2003.10.018.
14. Liu W, Zhang S, Hao L, Yang F, Jiao W, Li X, et al. Fabrication of carbon nanotubes/carbon fiber hybrid fiber in industrial scale by sizing process. *Appl Surf Sci* 2013;284:914–20. doi:10.1016/j.apsusc.2013.08.045.
15. Yuan H, Zhang S, Lu C, He S, An F. Improved interfacial adhesion in carbon fiber/polyether sulfone composites through an organic solvent-free polyamic acid sizing. *Appl Surf Sci* 2013;279:279–84. doi:10.1016/j.apsusc.2013.04.085.
16. Guigon M, Klinklin E. The interface and interphase in carbon fibre-reinforced composites. *Composites* 1994;25:534–9. doi:10.1016/0010-4361(94)90181-3.
17. Nass R, Schmidt H. Synthesis of an alumina coating from chelated aluminium alkoxides. *J Non Cryst Solids* 1990;121:329–33. doi:10.1016/0022-3093(90)90153-D.
18. Nabavi M, Doeuff S, Sanchez C, Livage J. Chemical modification of metal alkoxides by solvents: A way to control sol-gel chemistry. *J Non Cryst Solids* 1990;121:31–4. doi:10.1016/0022-3093(90)90099-8.
19. Shiba K, Onaka K, Ogawa M. Preparation of mono-dispersed titanium oxide–octadecylamine hybrid spherical particles in the submicron size range. *RSC Adv* 2012;2:1343–9. doi:10.1039/C1RA00748C.
20. Kikuchi M, Itoh S, Ichinose S, Shinomiya K, Tanaka J. Self-organization mechanism in a bone-like hydroxyapatite/collagen nanocomposite synthesized in vitro and its biological reaction in vivo. *Biomaterials* 2001;22:1705–11. doi:10.1016/S0142-9612(00)00305-7.
21. Mizushima Y, Ikoma T, Tanaka J, Hoshi K, Ishihara T, Ogawa Y, et al. Injectable porous hydroxyapatite microparticles as a new carrier for protein and lipophilic drugs. *J Control Release* 2006;110:260–5. doi:10.1016/j.jconrel.2005.09.051.

222   Advanced Manufacturing and Processing Technology

22. Park S-J, Seo M-K, Lee Y-S. Surface characteristics of fluorine-modified PAN-based carbon fibers. *Carbon N Y* 2003;41:723–30. doi:10.1016/S0008-6223(02)00384-6.
23. Jeong E, Kim J, Cho SH, Bae Y-S, Lee Y-S. Synergistic effects induced by oxy-fluorination of carbon preforms to improve the mechanical and thermal properties of carbon–carbon composites. *J Fluor Chem* 2011;132:291–7. doi:10.1016/j.jfluchem.2011.02.008.
24. Li CS, Vannabouathong C, Sprague S, Bhandari M. The use of carbon-fiber-reinforced (CFR) PEEK material in orthopedic implants: A systematic review. *Clin Med Insights Arthritis Musculoskelet Disord* 2015;8:CMAMD.S20354. doi:10.4137/CMAMD.S20354.
25. Petersen R. Carbon fiber biocompatibility for implants. *Fibers* 2016;4:1. doi:10.3390/fib4010001.
26. ASM International. Handbook Committee. n.d. ASM Handbook. Composite 2001; 21:1201. http://www.asminternational.org/search/-/journal_content/56/10192/06781G/PUBLICATION.
27. Massey T, Santacruz SR, Hou J, Pister KSJ, Carmena JM, Maharbiz MM. A high-density carbon fiber neural recording array technology. *J Neural Eng* 2018;16:016024.
28. Lee Y, Kong C, Chang JW, Jun SB. Carbon-fiber based microelectrode array embedded with a biodegradable silk support for in vivo neural recording. *J Korean Med Sci* 2019;34. doi:10.3346/jkms.2019.34.e24.
29. Ali MS, French TA, Hastings GW, Rae T, Rushton N, Ross ER, et al. Carbon fibre composite bone plates. Development, evaluation and early clinical experience. *J Bone Joint Surg Br* 1990;72:586–91.
30. Li Y, Wang D, Qin W, Jia H, Wu Y, Ma J, et al. Mechanical properties, hemocompatibility, cytotoxicity and systemic toxicity of carbon fibers/poly (ether-ether-ketone) composites with different fiber lengths as orthopedic implants. *J Biomater Sci Polym Ed* 2019;30:1709–24.
31. McCracken M, Lemons JE, Rahemtulla F, Prince CW, Feldman D. Bone response to titanium alloy implants placed in diabetic rats. *Int J Oral Maxillofac Implants* 2000;15:345–54.
32. Romanowska-Dixon B. Surgical management of choroidal melanoma. *Klin Oczna* 2005;107:635–41.
33. Blazewicz M. Carbon materials in the treatment of soft and hard tissue injuries. *Eur Cells Amd Mater* 2001;2. doi:10.22203/eCM.v002a03.
34. Schwitalla A, Müller W-D. PEEK dental implants: a review of the literature. *J Oral Implantol* 2013;39:743–9. doi:10.1563/AAID-JOI-D-11-00002.
35. Takahashi K, Tanabe K, Ohnuki M, Narita M, Ichisaka T, Tomoda K, et al. Induction of pluripotent stem cells from adult human fibroblasts by defined factors. *Cell* 2007;131:861–72. doi:10.1016/j.cell.2007.11.019.
36. Takahashi K, Okita K, Nakagawa M, Yamanaka S. Induction of pluripotent stem cells from fibroblast cultures. *Nat Protoc* 2007;2:3081–9. doi:10.1038/nprot.2007.418.
37. Komender J, Lewandowska-Szumieł M. Interaction between tissues and implantable materials. *Front Med Biol Eng* 2000;10:79–82. doi:10.1163/15685570052061937.
38. Price RL, Waid MC, Haberstroh KM, Webster TJ. Selective bone cell adhesion on formulations containing carbon nanofibers. *Biomaterials* 2003;24:1877–87. doi:10.1016/S0142-9612(02)00609-9.
39. Mäkisalo SE, Paavolainen P, Grönblad M, Holmström T. Tissue reactions around two alloplastic ligament substitute materials: experimental study on rats with carbon fibres and polypropylene. *Biomaterials* 1989;10:105–8. doi:10.1016/0142-9612(89)90041-0.
40. Minns RJ, Muckle DS. Mechanical and histological response of carbon fibre pads implanted in the rabbit patella. *Biomaterials* 1989;10:273–6. doi:10.1016/0142-9612(89)90105-1.
41. Kus WM, Gorecki A, Strzelczyk P, Swiader P. Carbon fiber scaffolds in the surgical treatment of cartilage lesions. *Ann Transplant* 1999;4:101–2. doi:http://dx.doi.org/.

Carbon Fibers

42. Tagusari O, Yamazaki K, Litwak P, Kojima A, Klein EC, Antaki JF, et al. Fine trabecularized carbon: ideal material and texture for percutaneous device system of permanent left ventricular assist device. *Artif Organs* 1998;22:481–7. doi:10.1046/j.1525-1594.1998.06152.x.

43. Kim C, Jeong YI, Ngoc BTN, Yang KS, Kojima M, Kim YA, et al. Synthesis and characterization of porous carbon nanofibers with hollow cores through the thermal treatment of electrospun copolymeric nanofiber webs. *Small* 2007;3:91–5. doi:10.1002/smll.200600243.

44. Aoki K, Usui Y, Narita N, Ogiwara N, Iashigaki N, Nakamura K, et al. A thin carbon-fiber web as a scaffold for bone-tissue regeneration. *Small* 2009;5:1540–6. doi:10.1002/smll.200801610.

45. Tiwari G, Tiwari R, Sriwastawa B, Bhati L, Pandey S, Pandey P, et al. Drug delivery systems: an updated review. *Int J Pharm Investig* 2012;2:2.

46. Schwartz M. *New Materials, Processes, and Methods Technology.* CRC Press; 2005.

47. Tangboriboon N. Carbon and carbon nanotube drug delivery and its characterization, properties, and applications. In *Nanocarriers Drug Delivery*, Elsevier; 2019, pp. 451–67.

48. Anliker U, Ward JA, Lukowicz P, Troster G, Dolveck F, Baer M, et al. AMON: a wearable multiparameter medical monitoring and alert system. *IEEE Trans Inf Technol Biomed* 2004;8:415–27.

49. Dias D, Paulo Silva Cunha J. Wearable health devices—vital sign monitoring, systems and technologies. Sensors 2018;18:2414.

50. Angelov GV, Nikolakov DP, Ruskova IN, Gieva EE, Spasova ML. *Healthcare Sensing and Monitoring*; 2019, pp. 226–62. doi:10.1007/978-3-030–10752-9_10.

51. Saccomandi P, Schena E, Caponero MA, Gassino R, Hernandez J, Perrone G, et al. Novel carbon fiber probe for temperature monitoring during thermal therapies. *2017 39th Annual International Conference of the IEEE Engineering in Medicine and Biology Society (EMBC)*, IEEE; 2017, pp. 873–6.

52. McConville A, Mathur A, Davis J. Palladium nanoneedles on carbon fiber: highly sensitive peroxide detection for biomedical and wearable sensor applications. *IEEE Sens J* 2018;19:34–8.

53. Bacáková L, Starý V, Kofronová O, Lisá V. Polishing and coating carbon fiber-reinforced carbon composites with a carbon-titanium layer enhances adhesion and growth of osteoblast-like MG63 cells and vascular smooth muscle cells in vitro. *J Biomed Mater Res* 2001;54:567–78.

54. Zhang D, Cabrera E, Zhao Y, Zhao Z, Castro JM, Lee LJ. Improved sand erosion resistance and mechanical properties of multifunctional carbon nanofiber nanopaper-enhanced fiber reinforced epoxy composites. *Adv Polym Technol* 2018;37:1878–85. doi:10.1002/adv.21846.

55. Xiong L, Zhan F, Liang H, Chen L, Lan D. Chemical grafting of nano-TiO_2 onto carbon fiber via thiol–ene click chemistry and its effect on the interfacial and mechanical properties of carbon fiber/epoxy composites. *J Mater Sci* 2018;53:2594–603. doi:10.1007/s10853-017-1739-5.

56. Mamourian AC, Mahadevan N, Reddy N, Marra SP, Weaver J. Prototypical metal/polymer hybrid cerebral aneurysm clip: in vitro testing for closing force, slippage, and computed tomography artifact. Laboratory investigation. *J Neurosurg* 2007;107:1198–204. doi:10.3171/JNS-07/12/1198.

57. Milavec H, Kellner C, Ravikumar N, Albers CE, Lerch T, Hoppe S, et al. First clinical experience with a carbon fibre reinforced PEEK composite plating system for anterior cervical discectomy and fusion. *J Funct Biomater* 2019;10:29. doi:10.3390/jfb10030029.

58. Xu Z, Zhang M, Gao S, Wang G, Zhang S, Luan J. Study on mechanical properties of unidirectional continuous carbon fiber-reinforced PEEK composites fabricated by the wrapped yarn method. *Polym Compos* 2019;40:56–69.

59. Rousseau M-A, Lazennec J-Y, Saillant G. Circumferential arthrodesis using PEEK cages at the lumbar spine. *J Spinal Disord Tech* 2007;20:278–81. doi:10.1097/01. bsd.0000211284.14143.63.

60. Rohner B, Wieling R, Magerl F, Schneider E, Steiner A. Performance of a composite flow moulded carbon fibre reinforced osteosynthesis plate. *Vet Comp Orthop Traumatol* 2005;18:175–82.

61. Godara A, Raabe D, Green S. The influence of sterilization processes on the micromechanical properties of carbon fiber-reinforced PEEK composites for bone implant applications. *Acta Biomater* 2007;3:209–20. doi:10.1016/j.actbio.2006.11.005.

62. Toldy A, Szolnoki B, Marosi G. Flame retardancy of fibre-reinforced epoxy resin composites for aerospace applications. *Polym Degrad Stab* 2011;96:371–6. doi:10.1016/j. polymdegradstab.2010.03.021.

63. Sell PJ, Prakash R, Hastings GW. Torsional moment to failure for carbon fibre polysulphone expandable rivets as compared with stainless steel screws for carbon fibre-reinforced epoxy fracture plate fixation. *Biomaterials* 1989;10:182–4. doi:10.1016/0142-9612(89)90021-5.

64. Bagheri ZS, Giles E, El Sawi I, Amleh A, Schemitsch EH, Zdero R, et al. Osteogenesis and cytotoxicity of a new carbon fiber/flax/epoxy composite material for bone fracture plate applications. *Mater Sci Eng C* 2015;46:435–42. doi:10.1016/j.msec.2014.10.042.

65. Petersen R. Titanium implant osseointegration problems with alternate solutions using epoxy/carbon-fiber-reinforced composite. *Metals (Basel)* 2014;4:549–69. doi:10.3390/met4040549.

66. Brochu ABW, Craig SL, Reichert WM. Self-healing biomaterials. *J Biomed Mater Res Part A* 2011;96A:492–506. doi:10.1002/jbm.a.32987.

Index

Abdulkareem, S. 12
abrasive grain size 73
abrasive jet micromachining (AJMM) method 74–75
  abrasive materials 76
  applications 78
  gas medium 77
  limitations 77
  MRR 77, 78
  nozzle 77
  process parameters 75–76
  working principle 75
abrasive slurry jet micromachining (ASJM) 114–115
accelerating voltage 99
acrylonitrile butadiene styrene (ABS) 149
actuation 186
adaptive control systems 9
adaptivity 186
additive manufacturing (AM) 148, 153
advanced micromachining methods 71
AGAEDM *see* argon gas-assisted EDM (AGAEDM)
Akram, J. 135
Aliakbari, E. 4
Aliyu, A. A. 13, 124
Alting, L. 113
aluminum-based MMC 114
Ambroziak, A. 138, 142
Amorim, F. L. 3
amplitude 73
analysis of variance (ANOVA) 4, 31, 39, 58
Ananthapadmanaban, D. 139, 141
Antonio et al. 142
Araujo, A. 4
argon gas-assisted EDM (AGAEDM) 5
Arivazhagan, N. 139, 141
Arooj, S. 7
Asif, M. 143
ASJM *see* abrasive slurry jet micromachining (ASJM)

Basak, A. 13, 124
Baseri, H. 3, 4, 10, 11
Bayramoglu, M. 7, 11
beam current 99
bearings 164
Bedheri et al. 219
Bennet, C. 143
Beri, N. 5
Bharathi, S. R. S. 140, 142

Bhatt, G. 10
Biermann, D. 53
Bifano, T. G. 117
binder jet technology (BJT) 150, 151
bioceramics 153
bioink 157
biomaterials 13, 152
biomimicry 154
bioscaffolding 155, 157
Bissacco, G. 121
Black, J. T. 114
Blazewicz, M. 213
bone–implant contact (BIC) 214
bone mineral content (BMC) 215
bone mineral density (BMD) 215
Bozkurt, B. 8
brass strip electrode 10
Buffa, G. 143

Calamaz, M. 55
Caliskan, S. 57
carbon fibers (CFs) 207
  biomedical sensor 216–217
  cobalt–nickel–chrome alloy 218
  composite 217–220
  composite implant for bone 213
  drug delivery 215–216
  mechanical properties 213
  nonemulsion 211
  phagocytosis 214
  regenerative medicine 214–215
  surface coating 210–212
  surface oxidation 208–210
  surface treatment methods 208, 209
  thermal treatments 212
  tissue growth 213–214
  XPS spectrum 210
CD *see* conventional drilling (CD)
Che-Haron, C. H. 54
chemical vapor deposition (CVD) 52
chip measurements 61
chip morphology 54, 60
clean operation 186
CN/TiAlN-coated carbide tools 57
coated tools 51–52
Cogun, C. 9, 12
continuous friction welding (CFW) 133
controlled capacity 186
conventional drilling (CD) 118
conventional micromachining 69–71, 103
copper-tungsten electrode 5

225

226 Index

cryogenically cooled copper electrode 8
cutting temperature 48
cylindrical rollers
surface 181
surface topography 180, 181

damping 186
Dang, X.-P. 4
deflection coils 98
designs of experimentation (DOEs) 31
desirability function analysis (DFA) 60
Dey, H. C. 135, 137
dielectric flow system 85
dielectric fluids 11–12
difficult-to-machine materials 50
Ding, S. 11
direct ink writing (DIW) 149
double-side lapping 164, 165; *see also* rollers
abrasive size, effect of 173, 174
downforce to roundness 175, 176
downforce to surface roughness 173, 174
experimental setup 172–173
MRR, rollers 175
drilling operations 46
Dornfeld, D. 111
Duffill, A. W. 7
Dursun, K. 9
dynamic area mill 24
dynamic contour 25
dynamic core mill 24, 25
dynamic recrystallization (DRX) 143
dynamic rest mill 25–26

EBMM *see* electron beam micromachining
(EBMM)
Egashira, K. 7
Ekmekci, B. 12
Ekmekci, N. 12
electrical discharge machining (EDM)
aluminum/alumina metal–matrix composite
11–12
biomaterials 13
CNC technology 11
dielectric fluid 11–12
disadvantages 2
experimental trials 10
history 84
layers 2
magnetic field-based electrical discharge
machining 9–10
multisparks 6
process parameters 3–4
pulse duration 7–8
selection of electrode 4–6
servo control mechanism 9
tools 10–11
tool wear 2–3

tool/workpiece 2
ultrasonic-assisted 8
vibratory tool/workpiece 8–9
electrochemical micromachining (ECMM)
advantages 82
applications 82
limitations 82
microgrooves 83
process parameters 79–81
removal rate model 81–82
surface finish 81
working principle 78–79
electrodischarge micromachining (EDMM)
advantages 86
applications 87
complex profiles 87
components 84–85
limitations 87
microholes 87
process parameters 85, 86
variants 85–86
working principle 83–84
electrodes 8, 79
electrode wear 3, 5
electrode wear rate (EWR) 5
electron beam micromachining (EBMM) 84
advantages 99
applications 100
cause/effect 98
components 96–98
drilling 100
equipment 96
history 95
limitations 99
material removal mechanism 96
microholes 101
pattern of holes 101
process parameters 98–99
schematic diagram 97
slot cutting 100
working principle 95, 96
electron gun 96
electronic and magnetic-based materials 112
elevated temperature machining 50–51
energy storage 186
Enomoto, T. 53
Eswara Krishna, M. 8
EWR *see* electrode wear rate (EWR)

fabrication 113
FDM *see* fused deposition modeling (FDM)
feed force 73
ferromagnetic shape memory alloy (FSMA) 187
field-emission scanning electron microscope
(FESEM) 194–196
flood cooling method 53
Floyd, R. W. 190

Index

focused ion beam technique 120
Fonda, P. 7
frame-type copper tools 11
frequency 73
friction coefficient, rollers
    abrasive concentration *vs.* friction coefficients
        168, 171
    Al_2O_3 abrasive slurry 169–170
    downforce *vs.* friction coefficients 170, 171
    lapping process 165–166
    polishing process 166
    SiC abrasive slurry 166, 168, 169
friction welding (FW)
    components 134
    definition 133
    ferrous metal alloys 135, 141
    ferrous/nonferrous metal alloys 134–135
    finite element model 142–143
    nonferrous metal alloys 141–142
    tensile properties 136–140
FSMA *see* ferromagnetic shape memory
    alloy (FSMA)
fused deposition modeling (FDM) 148

gas lasers 88–89
Gente, A. 54
Al-Ghamdi, K. 4
Ghoreishi, M. 7
Ghosh, A. 112
Godara, A. 218
Gopalakannan, S. 5
Gostimirovic. M. 7
Govindan, P. 10
Grant, B. 143
graphite 4, 6
graphite powder–added kerosene 12
grazing incidence X-ray diffraction (GIXRD)
    196–197
grey relational analysis (GRA) 3
gross domestic product (GDP) 46
Guo, C. 12

Hackert et al. 113
HA-EDM *see* hydroxyapatite-mixed
    electrical discharge machining
    (HA-EDM)
Hara, T. 190
Hascalik, A. 135, 139
Hayakawa, S. 9
Hazra, M. 141
heat-affected zone (HAZ) 141
heat-assisted machining 50–51
heating–cooling process 186–187
Heidenhain controller 27
helical chips, long tubular 58, 59
Helix angle 28–29
Hench, L. 153

high-energy regime 10
high-resolution transmission electron microscope
    (HRTEM) 197–200
high-speed machining (HSM)
    advantages 52
    benefits 24
    vs. conventional machining 24
    definition 24
    disadvantages 52–53
    element analysis, Ti-6Al-4V 55
    experimental setup
        DOE 31
        high helix cutter 28–29
        machine specifications 28
        machiningstrategy 29–30
        material selection 26
        parameters selection 30
        process parameters 28
        response surface methodology 31–33
        Ti-6Al-4V properties 26
    millingtoolpaths
        dynamic area mill 24
        dynamic contour 25
        dynamic core mill 24, 25
        dynamic rest mill 25–26
    results/discussion
        cycle time 37–41, 42
        spindle load 34–37, 42
    titanium alloy 55
Hoffmeister, H. W. 54
Holmes, M. 123
Hopkinson bar tests 55
hot machining 50–51
HRTEM *see* high-resolution transmission
    electron microscope (HRTEM)
HSM *see* high-speed machining (HSM)
Hua, J. 54
Huang, B. W. 120
Huang, J. T. 11
hydrocarbon oil 12
hydroxyapatite-mixed electrical discharge
    machining (HA-EDM) 13, 124

Inconel 718 6, 56, 57
independent self-assembly 154
industrialization 46
interelectrode gap (IEG) 79, 80
Islam, M. M. 5

Jadish et al. 4
James, J. A. 135
Jawaid, A. 54
Jayabharath, K. 135
Jedrasiak, P. 142
Jeswani, M. L. 12
Jian, L. 114
Jiang, R. 11

Kalpakjain, S. 113
Kimura, M. 134, 136
Klocke, F. 6
Kolli, M. 12
Kucukkose, M. 57
Kumar, A. 12
Kunieda, M. 6

Lakshminarayanan, A. K. 137, 142
laser-assisted machining 50–51
laser beam micromachining (LBMM)
    advantages 94
    applications 94–95
    beam intensity 91–92
    cause/effect 93
    laser ablation effect 90
    laser mask projection technique 91
    laser types 88
    limitations 94
    material removal 89–91
    micro/nanopulse duration lasers 91
    MMR 92–94
    schematic diagram 90
    working principle 88–89
laser mask projection technique 91
laser production crystal 88–89
lens current 99
Li, L. 12
Liang, Z. 136, 142
Lin, Y.-C. 12
linear friction welding (LFW) 143
Long, B. T. 11
low-energy regime 10
Lucca, D. A. 116
Luo, Y. 157
Lv, D. 118

machinability, judging
    chip form 48
    cutting temperature 48
    definition 46
    power consumption 47
    surface finish 47
    tool life 47
machining operation
    chip formation 54
    coated tools 51–52
    cutting tool 49
    flood cooling 53
    hot machining 50–51
    microgrooves 53–54
    MQL 51, 56
    superalloys 49, 50
magnesium-based MMC 114
magnetic field–based electrical discharge
        machining 9–10
magnetic lens 98

Maiman, T. H. 88
Maity, K. 57, 58, 59
Majumder, A. 3
Makisalo, S. E. 214
Mamalis, A. G. 10
manganese–vanadium tool steel workpiece 7
Manivannan, R. 4
Marafona, J. 3, 4
Marafona, J. D. 4
Markstedt, K. 157
martensite 186, 188, 190, 193
mask projection technique 92
Mastercam Dynamic Motion 29
Masuzawa, T. 123
material removal rate (MRR) 2, 69, 80, 144
Mathai, V. J. 10
mechanism simplicity 186
Mercan, S. 138, 141
Meshram, S. D. 138, 141
metal-matrix composite (MMC) 114
microdrilling 69
micro–electrodischarge grinding (μEDG) 86
microelectromechanical system (MEMS) 68,
        113, 187
microgrinding 69
microgrooves 53–54
microlubrication 51
micromachining process 104, 113
    applications 112
    characteristics 114
    classification 69
    cutting energy 115–116
    cutting fluid 120–122
    ductile modes 116–117
    edge/surface finishing 117–119
    machines/tools/systems 119–120
    machine tool components 122–123
    metrology 123–125
    orthogonal flying cutting 115, 116
    process physics 113–115
    sectioned hole 117, 118
    vs. time 112
    worpiece/design issues 119
micromilling 69, 70
microtexturing 61
microturning 69, 70
microwire EDM 85
middle-energy regime 10
milling experiments 55
milling operation 46
miniature tissue blocks 155
miniaturization 112, 113, 114; *see also*
        micromachining process
minimum quantity of lubricant (MQL) 51, 56,
        121
Miyoshi, T. 123
Mizumoto, H. 122

Index

MMC *see* metal-matrix composite (MMC)
Mohri, N. 3, 6
Morris et al. 213
MRR *see* material removal rate (MRR)
MT-CVD-coated cutting tool 59–60
multifunctionality 187
multiobjective optimization on the basis of ratio
        analysis (MOORA) 57

Naidu, K. 10
nanohydroxyapatite (nHA) 13, 124
"near dry lubrication" 51
Neuro-Grey method 4
Ng, P. S. 12
nickel–titanium (NiTi) system
    benefits 202–204
    conversion pathways 191–192
    invariant reaction points 191
    metallurgical factors 191
    Ni52Ti48 alloy 192
    nonuniformity 193–194
    physical metallurgy
        alloys 190
        martensitic transformation 191–194
        phase diagrams 188–191
    physical properties
        FESEM 194–196
        GIXRD technology 196–197
        HRTEM 197–200
    stress-strain temperature (α-ε-T) curve
        187–188
    three-phase transformation 192, 193–194
Nitinol 186, 187
Nitinol Naval Ordnance Laboratory 186
noiseless/spark-free operation 186
nonconventional micromachining
    AJMM 74–75
    EBMM 95–100
    ECMM 78–83
    EDMM 83–87
    LBMM 88–95
    PAMM 100–102
    USMM 71
nontraditional machining methods 71
normal copper electrode 8
Nouraei, H. 114

Obikawa, T. 55
Okka, M. A. 6
one-factor-at-a-time (OFAT) 31
operating energy 186
optimization process 56
organs-on-chips model 155
orthogonal flying cutting 115, 116

Pan, L. 140, 142
Panda, D. K. 4

Parida, A. K. 57
particle swarm optimization (PSO) 3
Patowari, P. K. 7, 8
PCA *see* principal component analysis (PCA)
PEEK *see* polyether ether ketone (PEEK)
Pey Tee, K. T. 8
phagocytosis 214
PLA *see* polylactic acid (PLA)
plasma arc micromachining (PAMM)
    advantages 102
    applications 102
    limitations 102
    process parameter 101–102
    working principle 101
plasma-assisted machining 50–51
plunger-type water pump 115
polishing process
    abrasive size, surface roughness 177–178
    downforce
        MRR 178
        surface roughness 178, 179
        to roundness 180–182
        vs. roundness 181, 182, 183
    experimental setup 177
    measured profiles 181
    parameters 177
polyether ether ketone (PEEK) 153, 214
polylactic acid (PLA) 149
postweld heat treatment (PWHT) 135
power density 99
powder metallurgical (PM) steel 135
powder-mixed-EDM 86
power to volume ratios 186
Pradeep Kumar, M. 4, 5
Pradhan, S. 13, 56, 57, 58, 59, 61
Pragadish, N. 5
Prakash, C. 13, 124
Prasanthi, T. N. 136, 141
principal component analysis (PCA) 4
projection stereolithography (pSLA) 148, 149
PSO *see* particle swarm optimization (PSO)
pulse discrimination methods 8
pulse durations 99
pulse generation 84
pulse wave 8
PVD Al-Ti-N coating carbide, KC5010 58
PWHT *see* postweld heat treatment (PWHT)

Rajput, R. K. 112
Rajurkar, K. P. 9
rapid prototyping technique 148
recovery 186
reliability 187
response surface methodology (RSM) 31, 58
reverse micro-EDM 86
RFW *see* rotary friction welding (RFW)
rhombohedral 190

Index

Ribeiro, M. V. 54
Rohner, B. 218
rollers
  abrasive concentration *vs.* friction coefficients 168, 171
  downforce
    to roundness 175, 176
    to surface roughness 173, 174
    vs. friction coefficients 170, 171
  lapping parameters 167
  measured profiles 176
  MRR 175
  polish (*see* polishing process)
  polishing parameters 168
rotary friction welding (RFW) 133
rotary tool 5
rotary ultrasonic drilling (RUD) 118
rotary ultrasonic machining (RUM) 117
RSM *see* response surface methodology (RSM)
Ruby laser 88

Saburi, T. 190
Sarsilmaz, F. 135, 140
scanning electron microscope (SEM) 103
Schaller, T. 119
Schwitalla, A. 214
Seli, H. 135, 140
Sell, P. J. 219
Selvamani, S. T. 141
Sen, I. 6
sensing 186
sensor application *vs.* level of precision 124
Seo, Y. W. 116
servo control mechanism 9
servo control system 85
Shabgard, M. R. 3, 8
Shankar, P. 8
shape memory alloy (SMA) 185
  biomedical equipment 202
  microgripper and microtweezer 202
  microvalves/micropumps 201–202
  NiTi (*see* nickel–titanium (NiTi) system)
  properties 186–187
shape memory effect (SME) 185
Sharma, A. 119, 123
Shervani-Tabar, M. T. 9
Shiba, K. 211
silica-based bioglasses 153
Singh, A. K. 12
Singh, N. K. 5
Singh, R. P. 8
Singhal, S. 8
Singh Bains, P. 10
Sohani, M. S. 6
solid-state lasers 88
Son, S. M. 7

Song, K. Y. 10
Soni, J. S. 2
spark erosion mechanism 84
Spritam® 150
Srivastava, V. 8
Srinivasan, M. 140, 142
stainless steel (SS) 49, 135
Sugihara, T. 53
Sun, S. 55
superalloys 49, 50
surface roughness (SR) 4
Suzuki, K. 6

Tabari, C. 7
Taguchi method 4, 7, 8, 56, 57, 58, 59
Tangboriboon, N. 215
Taniguchi, N. 111
El-Taweel, T. A. 6
Taylan, O. 4
Taylor, A. 190
Teimouri, R. 3, 10
textured cutting tools 62
thermal cycling 193
thermoelectric gun 97
3D bioprinting
  approaches 154
  biomimicry 154
  independent self-assembly 154–155
  issue scaffold 156
  miniature tissue blocks 155
  tissue repair 156
three-dimensional printing (3DP) 148
  biomedical applications 149
  components 152
  fabrication 148
  medicine
    bioscaffolding 155–157
    materials 152–153
    scaffolds fabricated 152, 153
    in situ bioprinting (*see* 3D bioprinting)
  methods 148
  pharmaceuticals 150
  powder jet technology 151
  shapes/sizes 150
3D simulation, Mastercam 30
thermally enhanced machining 50–51
titanium bar 27
tool flank wear 60
tool life 47
tool wear 2–3
tool wear rate (TWR) 2
tool–workpiece interface 3
Toren, M. 5
Torres, A. 5
TOSIS method 4
transmission electron microscope (TEM) 103

Index

Tripathy, D. K. 4
Tripathy, S. 4

Uday, M. B. 137, 140
Udayakumar, T. 140, 141
Ueda, K. 117
ultrasonic-assisted cryogenically cooled copper electrode 8
ultrasonic-assisted EDM (US-EDM) 8
ultrasonic micromachining (USMM)
    abrasive slurry 72
    advantages 74
    applications 74
    experimental setup 73
    feed mechanism 72
    limitations 74
    MRR 73–74
    oscillating system 72–73
    process parameters 73
    tool material 71–72
    working principle 71, 72, 73
Umbrello, D. 55
Unses, E. 12

vacuum chamber 97–98
Vairamani, G. 135
vertical machining center (VMC) 26, 41
Vinothkumar, S. 3
Von Turkovich, B. F. 114

Wang, J. H. 11
Wei, Y. 136, 141
Winiczenko, R. 134
Won, S. 140, 142

Yan, B. H. 7
Yilmaz, O. 6

Zanger, F. 55, 56
Zeng et al. 211
Zetek, M. 56
Zhan et al. 57
Zhang, Q. H. 8, 12
Zhang, S. 54, 56, 57
Zhang, Y. 12
Zhong, Z. W. 116